Possible Causes of the Environmental Kuznets Curve

European University Studies
Europäische Hochschulschriften
Publications Universitaires Européennes

Series V
Economics and Management

Reihe V Série V
Volks- und Betriebswirtschaft
Sciences économiques

Vol./Band 3040

PETER LANG
Bern · Berlin · Bruxelles · Frankfurt a. M. · New York · Oxford · Wien

Christoph Lieb

Possible Causes of the Environmental Kuznets Curve

A Theoretical Analysis

PETER LANG
Bern · Berlin · Bruxelles · Frankfurt a. M. · New York · Oxford · Wien

Bibliographic information published by Die Deutsche Bibliothek
Die Deutsche Bibliothek lists this publication in the Deutsche Nationalbibliografie; detailed bibliographic data is available on the Internet at ‹http://dnb.ddb.de›.

British Library and Library of Congress Cataloguing-in-Publication Data:
A catalogue record for this book is available from *The British Library*, Great Britain, and from *The Library of Congress*, USA

Doctoral dissertation approved by the Wirtschaftswissenschaftliche Fakultät, University of Heidelberg, 5th February 2002.

ISSN 0531-7339
ISBN 3-03910-338-5
US-ISBN 0-8204-7010-4

Hochfeldstrasse 32, Postfach 746, CH-3000 Bern 9, Switzerland
info@peterlang.com, www.peterlang.com, www.peterlang.net

Contents

Meiner Familie
und Saskia

Danksagung

An allererster Stelle bin ich meinem Lehrer und Doktorvater, Prof. Dr. Till Requate, zu grossem Dank verpflichtet. Er hat mehrere Fassungen meiner Dissertation durchgelesen und mir jedes Mal äusserst hilfreiche Anregungen gegeben, die es mir ermöglichten, sie sprachlich und vor allem inhaltlich zu verbessern. Hätte er nicht geraten, ich solle ordinale Nutzenfunktionen benutzen, so hätte ich mir wohl nie über ordinale und kardinale Nutzenfunktionen Gedanken gemacht und Kapitel 4 und 7 meiner Dissertation wären nicht geschrieben worden.

Mein Dank gilt auch meinem Zweitgutachter, Prof. Dr. Clive Bell. Auch er hat frühere Fassungen meiner Dissertation durchgelesen und mir wesentliche und überaus nützliche Hinweise und Ansporn gegeben. Bei Prof. Dr. Malte Faber möchte ich mich bedanken für mehrere höchst interessante Gespräche, die mich immer wieder ungemein motiviert haben.

Hilfreiche Kommentare habe ich auch von einigen Kollegen und Gästen an der Universität Heidelberg bekommen, die Teile meiner Dissertation gelesen haben. Es sind dies Dr. Bouwe Dijkstra, Ulf Moslener, Dr. Sjak Smulders (Katholieke Universiteit Brabant, Tilburg), Dr. Sander de Bruyn (Vrije Universiteit Amsterdam), Prof. Dr. John List (University of Arizona) und Dr. Giuseppe di Vita (Università di Catania). Von zwei anonymen Gutachtern der Zeitschrift *Environment and Development Economics* habe ich ebenfalls wertvolle Anregungen empfangen. In dieser Zeitschrift ist eine überarbeitete Fassung von Kapitel 5 veröffentlicht worden (Nr. 7 (2002), S. 429–448).

Nützliche Hinweise habe ich auch erhalten in vielen Gesprächen mit meinen Kollegen Olaf Hölzer, Dagmar Nelissen, Martin Quaas, Dr. Andreas Lange, Patrick Graichen, Oliver Kopp, PD Dr. Frank Jöst, Dr. Stefan Baumgärtner, Henrike Koschel, Annegret Steinmetz und Harald Tauchmann. Ihnen allen sowie Christian Becker, Jörg Breitscheidel, Mariah Khovanskaia, Susanne Klimpel, Andreas Löschel, Paul Mensink, Marian Mraz, Johannes Schiller und Ralph Winkler gilt auch mein Dank für die gute Stimmung, die immer im Graduieretenkolleg und am interdisziplinären Institut für Umweltökonomie geherrscht hat.

Für die finanzielle Unterstützung während den letzten drei Jahren bin ich dem Graduiertenkolleg 'Umwelt- und Ressourcenökonomik' dankbar, das von der Deutschen Forschungsgemeinschaft getragen wird. Das Graduiertenkolleg und sein Sprecher, Prof. Dr. Till Requate, ermöglichten es mir auch, Teile meiner Dissertation an internationalen Konferenzen vorzutragen.

Schliesslich möchte ich mich auch bei meinen Eltern und bei meinem Bruder bedanken, die mich schon seit meiner frühen Kindheit immer wieder zum Denken anregten und somit auch zum Entstehen dieser Arbeit beigetragen haben. Mein herzlichster Dank jedoch gilt Saskia: Sie hat mir die schönsten drei Jahre meines Lebens geschenkt.

Heidelberg Christoph Lieb

1. Introduction

1.1 The environmental Kuznets curve

In 1991 *Grossman* and *Krueger* found evidence that some pollutants are rising with income at low income levels, but at a higher income level a turning point is reached and further income growth subsequently leads to lower pollution (see figure 1.1). *Panayotou* (1995, first published 1993) called this inverted U-shaped pattern an environmental Kuznets curve (EKC) after the original Kuznets curve which describes an inverted U-shaped relationship between income inequality and income (*Kuznets* 1955). Both the EKC and the original Kuznets curve are based on the idea that in the course of economic growth some measure of the quality of life – environmental quality or an equal income distribution – initially deteriorates before improving again (*Pearson* 1994, 201 and *Munasinghe* 1999, 95). More generally, the behaviour of pollution when income rises can be called a pollution-income relationship (PIR). Thus the EKC is a special case of the PIR – namely an inverted-U shaped PIR.

Many environmentalists believe that economic growth necessarily involves a degradation of environmental quality (*Cole* 1999, 87), i.e. they believe that the PIR is monotonically rising. If this is the case, growth must sooner or later come to a halt since the world's capacity to absorb pollution is only limited (*López* 1994) – unless there is pollution-saving technological progress which allows production to increase without causing additional pollution (*Stokey* 1998). Hence, in this view solving the pollution problem is one of the main challenges of the 21st century (*Homburg* and *Mathies* 1998, 13).

However, the finding of an EKC by *Grossman* and *Krueger* (1991) and subsequently by many other authors has given rise to

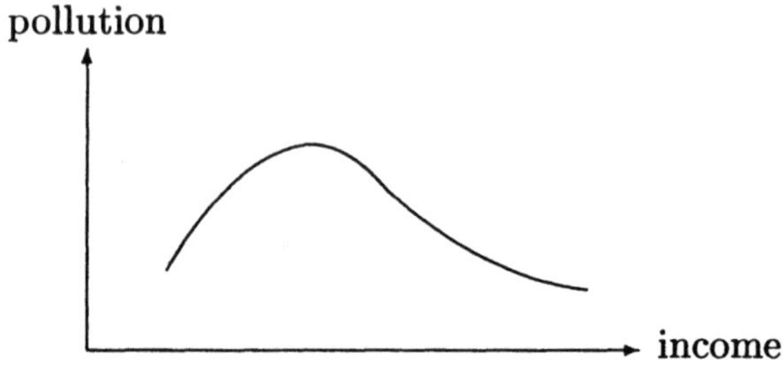

Figure 1.1: The environmental Kuznets curve.

a completely different opinion: Pollution is considered as only a transitional phenomenon in the course of economic growth. In this view 'in the longer run, the surest way to improve your environment is to become rich' (*Beckerman* 1992, 491). Hence, 'most governments and global institutions [...] see no conflict between economic growth [and environmental degradation]' because they 'typically argue that only economic growth can provide the resources with which to tackle environmental problems' (*Cole* 1999, 91 and 87). *Beckerman* (1992, 482) even jumps to the conclusion that 'there is clear evidence that [...] the best – and probably the only – way to attain a decent environment [...] is to become rich'. However, such 'clear' evidence does not exist.

In fact, before the first EKC studies appeared there was hardly any evidence concerning the relationship between economic growth and the environment. Therefore we will not discuss the often surveyed literature on growth and the environment published before the EKC studies showed up. A survey of this literature can, for instance, be found in *Cole* (2000b, chapter 4) or *de Bruyn* (2000, part I). Furthermore, the recent EKC evidence has been derived with reduced form models. These models only show what shape the PIR has, but they do not allow to draw any conclusions why that shape arises. Moreover, the evidence suggests that the PIR is only an EKC for local and flow pollutants whereas the PIR is monotonically rising for global and stock pollutants. Therefore the optimistic interpretations – which led 'policy makers such as the World Bank and the General Agreement on Tariffs and Trade (GATT) [to] use the results as evidence that economic growth should be promoted as a method of improving the environment' (*Kelly* 2000, 2) – might be wrong.

The main purpose of this book is to answer the question why an EKC arises. In other words, we want to find possible causes of the EKC. Put differently, we are looking for explanations why there is an EKC for some pollutants, but not for others. Many different possible causes of the EKC have been proposed in the literature. It is impossible to say which of these possible explanations are important without looking more closely at some further evidence. Therefore part I of this book gives the first survey of all the different possible causes of the EKC which have been proposed in the literature and reviews the corresponding evidence. The main conclusion is that the above optimistic interpretation of the EKC is not tenable. *Ayres* (1995, 97) even states that this interpretation is 'false and pernicious nonsense'. Instead, environmental policy is responsible for the downturn of the EKC. If there is no regulation, however, the PIR is monotonically rising.

In part II and III we analyse one of the causes of the EKC more closely and we offer two new explanations for the EKC. In part II rising demand for environmental quality, one of the causes which have been advanced, is theoretically scrutinized. We show that a tendency to satiation in consumption causes demand to rise. Furthermore, an EKC might only exist if the elasticity of intertemporal substitution is sufficiently high, i.e. if a smooth development path is not of high priority. In part III, finally, we prove that ever rising stock pollution and the neglect of future damages might cause the downturn of the EKC for a flow pollutant.

1.2 An overview of the book

Part I of the book reviews the EKC literature. First, the empirical literature is summarized in chapter 2. This literature survey improves upon previous work because it also deals with several conflicting results published in the last few years. This implies that we will draw some different conclusions. After analysing the estimation technique, the results derived with this method are presented: An EKC arises for flow and local pollutants, but the PIR is monotonically rising for stock and global pollutants as well as for aggregate measures of pollution. We then critically examine the estimation technique which is used to derive these results. We show that pollution data are of poor quality, that data are even missing for many types of pollution, that the use of a time trend can be problematic, that multicollinearity might hamper the results – which has been neglected by the literature to date – that different countries have different PIRs, and that shocks might have permanent effects on the PIR. Finally, some emission forecasts show that emissions will continue to rise fast. The main conclusion from chapter 2 is that the EKC exists, but only for some pollutants and not in all countries alike.

Chapter 3 gathers all the different possible causes of the EKC which have been proposed in the literature. On the one hand, we discuss various theoretical models and arguments which have been put forward to explain the EKC. On the other hand, we review the empirical literature to see whether or not these explanations are supported by the data. Such a detailed survey of all possible causes of the EKC and their empirical validity has never been given before. The rising branch of the EKC can be explained by the fact that *ceteris paribus* more production causes more pollution. The downturn of the EKC occurs since the *ceteris paribus*-condition is violated.

Pollution declines because of one of the following reasons: Because demand for environmental quality rises with income (possibly because of the fear of incurring an irreversible damage) which induces stricter environmental policies, because one pollutant is reduced by substituting it by another pollutant, because technological progress allows to produce more while causing less pollution, or because increasing returns to scale in abatement allow richer countries to abate at lower average costs. These causes are supported by the empirical evidence. Only partly empirically validated is the hypothesis that the EKC arises because dirty industries migrate out of developed countries into developing countries. Finally, the effects of structural change and of the income distribution on the EKC are only small or even insignificant. It is also shown in chapter 3 that a falling EKC does not necessarily imply that environmental quality is improving because irreversible damages might have happened. The main conclusion of this chapter is that an EKC emerges when policy measures are taken, but that the PIR is expected to be monotonically rising when the government remains inactive.

In part II of the book we analyse why demand for environmental quality rises with income. Before doing so, however, we consider some methodological issues in chapter 4. We show that in many static models there are assumptions about the cardinal utility function. However, a static model can be solved with milder assumptions about the ordinal utility function only. We offer a method how this can be done. The method has four additional advantages: It ensures that all goods are normal goods, it prevents interpretations which violate the normal good assumption, it allows to write equations in a more compact manner, and it can easily be interpreted graphically. We also consider dynamic models in chapter 4. We show that a monotonic transformation of the felicity function (also called the instantaneous utility function) is an ordinal change: The elasticity of intertemporal substitution is altered. Hence, we find a method how to check which results are robust to changes in the elasticity of intertemporal substitution: We can analyse which results are robust to monotonic transformations of the felicity function. We apply these two methods in chapters 5–8.

In chapters 5–7 we scrutinize one of the empirically validated causes of the EKC: Rising demand for environmental quality. We show that a tendency to satiation in consumption can cause demand for environmental quality to rise. To do so we reconsider the simple static model of *McConnell* (1997) in chapter 5. We relax several of his assumptions and we draw different conclusions from the model. In the model consumption and environmental quality are both normal goods. Pollution is generated by consumption and can be abated. We show that at low income levels no abatement is optimal and pollution increases with income. Once abatement expenditures are positive, we demonstrate that satiation in consumption is not only sufficient to find an EKC, but a tendency to satiation is even necessary if we assume a standard functional form for the pollution function. In other words, for standard functional forms it is necessary for an EKC to exist that environmental quality is a normal good. We also verify that – in contrast to different claims in the literature – the relationship between the income elasticity of demand for environmental quality and the EKC is ambiguous. Finally, we prove that the model of *Stokey* (1998) – which at first glance seems rather different from our model – is actually an interesting special case of it. We offer a new interpretation of *Stokey's* (1998) result: With asymptotic satiation there is an EKC, but without satiation pollution is monotonically rising.

In chapter 6 we endogenize growth. The static model in chapter 5 allows us to work out some conditions for the EKC by simple comparative statics. A model of the EKC, however, should be dynamic as the EKC is an inherently dynamic phenomenon in the course of economic growth. To endogenize growth the standard Ramsey approach is applied. Surprisingly, we find almost the same results as in the static model. Furthermore, almost all results are robust to changes in the elasticity of intertemporal substitution. In particular, the shape of the PIR – i.e. whether it is monotonically rising or an EKC – is robust. Moreover, if there is an EKC, the pollution level at the turning point is also robust. Only the income level at the turning point rises with the elasticity of intertemporal substitution.

In chapter 7 we offer a first new explanation of the EKC. We study the same model as in chapter 6 except that we assume that pollution is generated by production capital instead of consumption. Thus we reconsider and extend the model of *Selden* and *Song* (1995). We find that the steady state pollution level is zero in chapter 7 under almost the same conditions as in chapter 6. Then the PIR is an EKC. However, the income and pollution level at the turning point of the EKC are both high when the elasticity of intertemporal substitution is high. The elasticity of intertemporal substitution is even more important when the steady state pollution level is positive: Then an EKC exists only if the elasticity of intertemporal substitution is sufficiently high. Otherwise, the PIR is monotonically rising. Intuitively, when the elasticity of intertemporal substitution is small, a smooth development path is preferred. Thus we do not accept the high pollution levels around the turning point of the EKC. Instead we prefer a smooth, monotonically rising development of pollution. Hence, if pollution is generated by production capital, the elasticity of intertemporal substitution is crucial for the shape of the PIR.

In part III we supply another new explanation of the EKC: The EKC for a flow pollutant might be due to the neglect of future damages and due to ever rising stock pollution. The main purpose of part III is to find a possible explanation of the empirical evidence that the PIR is an EKC for flow pollutants, but that the PIR is monotonically rising for stock pollutants. We use an overlapping generations model based on the model of *John* and *Pecchenino* (1994), and we extend it to two pollutants. To my knowledge this is the first model of the EKC with two pollutants. The flow pollutant causes immediate damages, but the stock pollutant harms the environment only in the future.

In chapter 8 we examine the solution of a succession of myopic governments and in chapter 9 we sketch the far-sighted optimum. We show that myopic governments never abate stock pollution. Instead they abate flow pollution. Therefore they follow a path on which flow pollution exhibits an EKC, but on which stock pollution rises monotonically. This is consistent with the empirical evidence. Without the stock pollutant, however, the PIR of the flow pollutant can be monotonically rising. Thus the growth of the stock pollutant can actually cause the EKC for the flow pollutant. Furthermore, we find that the turning point of the EKC for the flow pollutant depends on the level of stock and flow pollution.

In chapter 9 we consider the far-sightedly optimal path of the same model as in chapter 8. Under certain parameter constellations far-sighted governments only abate stock pollution while myopic governments never do so. These constellations are more probable when abatement of the flow pollutant increases stock pollution which is often observed in reality. In these cases flow pollution increases monotonically with income and the myopically optimal EKC is due to the neglect of future damages. However, it is possible that an EKC for the flow pollutant is also optimal for far-sighted governments. Finally, we argue that it is impossible to internalize all external effects because of coordination problems and transaction costs and because some damages are difficult to measure. Therefore the myopic solution – which is consistent with the empirical evidence – might be closer to reality than the far-sighted solution. Hence, we conclude that an EKC can only be expected for some (flow) pollutants, but that the PIR of other (stock) pollutants tends to be monotonically rising.

In chapter 10, finally, we draw the main conclusions from our analysis. We summarize what possible causes of the EKC we have found and why there is an EKC for some pollutants, but not for others.

Part I

The Environmental Kuznets Curve: A Survey of the Literature

2. The Empirical Literature

2.1 Introduction

In part I of the book we survey the literature dealing with the EKC. This survey is divided into two chapters. In chapter 2 we first critically analyse the empirical evidence of the PIR to see whether or not an EKC actually exists or for which pollutants it exists. In chapter 3 we examine all the possible causes of the EKC which have been proposed in theoretical models or arguments. In that chapter we also scrutinize whether or not these explanations of the EKC are supported by empirical evidence.

The survey of the empirical literature improves upon previous work because it also considers several conflicting results published in the last few years. Therefore we will draw some different conclusions.

This chapter is organized as follows. In section 2.2 we describe the data situation and the pollutants for which the PIR has been estimated. In section 2.3 we review the traditional empirical EKC studies which all employ a cross country analysis. First, the traditional cross country estimation technique is discussed in subsection 2.3.1, then the results for several pollutants and for some aggregate measures of pollution are presented in subsection 2.3.2, and finally the traditional technique is criticized in subsection 2.3.3. This criticism leads to the conclusion that time series analysis is more appropriate. Thus in section 2.4 results from time series analyses are surveyed. In section 2.5 we give an overview of some emission forecasts. Section 2.6 concludes.

2.2 Data and pollutants

One of the main problems when estimating the PIR is finding reliable and comparable data for pollution (*Millimet* 2000, 2). In fact, the lack of large and standardized pollution data sets is notorious (*Ansuategi* 2000, 43). In different countries data are often measured by different methods and possibly at unrepresentative locations (*Shafik* 1994). Furthermore, sufficiently long time series are missing because collection of environmental data has started in the 1960s at the earliest, but for many countries and pollutants even much later. Data from developing countries are even more rare and 'often considered unreliable' (*Auffhammer et al.* 2001a, 3). Moreover, for many dimensions of environmental quality data are missing such that the PIR cannot be estimated for these dimensions. Hence, most studies focus on those pollutants for which data are available, i.e. on air and water pollutants.

We now give an overview of the pollutants which have been analysed in the empirical literature and of some data related issues concerning these pollutants. For a survey of the exact data sources for all the pollutants discussed below we refer to *Ansuategi* (2000, 32–35). Air pollutants such as sulphur dioxide (SO_2), suspended particulate matter (SPM), oxides of nitrogen (NO_x), carbon monoxide (CO), lead, and volatile organic compounds (VOC) are all connected with important adverse health consequences (*Selden* and *Song* 1994, 150, *Grossman* 1995, 29, and *Carson et al.* 1997, 438). Because of two incomparable methods of measurement some researchers estimate two regressions for SPM: One method indicates dark material in the air (dark matter), the other measures the total weight of SPM (heavy particles) (*Grossman* and *Krueger* 1991, 10). The data situation for the most prominent greenhouse gas, carbon dioxide (CO_2), is the best of all pollutants as relatively long time series are available for many developed and developing countries (*Roberts* and *Grimes* 1997, 192). However, for CO_2 only emission data are available.

For the other indicators of air pollution there has been a discussion over the advantages and disadvantages of emission per capita and concentration data. Emissions are directly connected with current economic activity, but they are independent of the area in which they are released and therefore of the damage they cause. Concentrations, in contrast, measure the local impact on the environment, but they are not directly related to local, current economic activity – especially when the pollutant is transboundary and long-lived (*Ansuategi* 2000, 82).[1]

We might ask whether the income level at the turning point of the EKC should be expected to be higher for emissions or for concentrations. Note first that concentration data are mostly gathered in cities (*Stern* 1998, 182) while emission data are on the country level. *Selden* and *Song* (1994, 148) argue that the turning point of the EKC should lie at a higher income level for country emissions than for urban concentrations because urban concentrations can easily be reduced by using higher smoke stacks or by moving dirty industries out of urban areas and because urban air quality has immediate human health effects. Another effect has been overlooked in the EKC literature: Emissions might be falling, but might still lie above the assimilative capacity of nature. Then the concentration continues to increase (cf. figure 3.4 on page 87). Only when emissions fall below the assimilative capacity, the concentration also starts to decline. In this case the turning point for concentrations lies at a higher income level than for emissions.[2] The evidence tends to support the first argument – namely that the turning point for country emissions lies at a higher income level – (*Cavlovic et al.* 2000), but the evidence is ambiguous.

Furthermore, emission and concentration data both have their measurement problems. Emission data are not measured, but

[1] *Ansuategi* (2000, chapter 5) therefore proposes to use emissions per area which measure the potential damage irrespective of where it occurs. He obtains similar results than other authors with emissions per capita or concentrations.

[2] *Kelly* (2000, figures 6 and 7) finds this result in his model, but he does not point it out.

calculated from estimates of fuel use and emission coefficients for different fuel types (*de Bruyn* 2000, 100). Concentrations vary widely over time and space. Moreover, the monitoring stations are often located at sites where pollution is considered an actual or potential problem (*Grossman* 1995, 24). Due to this site selection bias we tend to overestimate concentrations (*Borghesi* 1999). Even worse, the setting up of a monitoring station shows that there is some political will to combat pollution. This might bias the results towards the EKC: When the monitoring station is set up, concentrations might be high and rising. Thus pollution is tackled and concentrations subsequently fall. Nevertheless, *Kaufmann et al.* (1998, 211) conclude that concentration data are more reliable. However, since neither emissions nor concentrations are clearly superior, it is advisable to estimate the PIR with both data sets.

In addition to the estimations for air pollutants, the PIR has also been studied for three groups of river water quality indicators measured as concentrations: First, the state of the oxygen regime (dissolved oxygen (DO), biological oxygen demand (BOD), chemical oxygen demand (COD), and nitrates) which is crucial for the fish population, second, the concentration of pathogens (fecal and total coliform (FCOL and TCOL)) which cause a variety of sometimes fatal diseases, and, third, the concentration of heavy metals (lead, cadmium, arsenic, mercury, and nickel) which bioaccumulate in fish and – when eaten by man – result in many health risks (*Grossman* and *Krueger* 1995, 357–359).

The relationships between the lack of clean drinking water and income and between the lack of urban sanitation and income have also been analysed. Strictly speaking, however, these are not pollution indicators, but measures of the lack of basic services (*Ekins* 1997, 811). Moreover, the PIR has also been estimated for municipal waste and for deforestation. Deforestation data are notoriously unreliable due to three reasons: First, because they are often estimated from population growth rates, second, because countries which have already cut down most of their original forests seem to have undergone an improvement (*Shafik* and *Bandyopadhyay* 1992), and, third, because there are important differences in biodiversity between primeval forests and monocultural plantations which are lost in the data. Finally, there are also studies examining the PIR for some aggregate measures of pollution. These measures will be considered in paragraph 2.3.2.1 below.

Hence, the PIR has been estimated for many pollutants. Many other indicators of pollution, however, are difficult to measure, in particular soil erosion (*van Kooten* and *Bulte* 2000, 387), desertification, pollution and depletion of groundwater aquifers, biodiversity loss (*Cole* 1999, 95), acidification (*Neumayer* 1998, 168), extinction of animal and plant species, climate change, the ozone hole, and the risk of nuclear power stations. *Karr* and *Thomas* (1996) argue that EKC results obtained for the human health related environmental indicators discussed above should not be considered as representative for overall environmental quality or ecological health. Similarly, *Arrow et al.* (1995) state that our main interest should lie in ecosystem resilience. Resilience is a measure of the magnitude of shocks an ecosystem can accommodate and still return to the same equilibrium as before. If pollution oversteps a certain threshold level, in contrast, the ecosystem flips into another, possibly environmentally disastrous equilibrium. So we should measure pollution relative to this threshold. Furthermore, environmental quality is also threatened by so-called chemical time bombs: These are chemicals which first accumulate without causing any harm, but are then suddenly set free and damage the environment (*Stigliani et al.* 1991). However, neither resilience (*Ayres* 1995, *Schindler* 1996, and *Orians* 1996) nor chemical time bombs are measurable.

Summarizing we conclude that the data quality is poor and that data are only available for recognized, potentially highly damaging, and easily measurable pollutants. In fact, the choice of a pollution indicator is often determined by data availability (*de*

Bruyn 2000, 73). Hence, even if we found that there is an EKC for all the pollutants which have been analysed, we could not conclude that there is an EKC for overall pollution.

To estimate the PIR we need not only pollution data, but also income data. Almost all studies use gross domestic product (GDP) in purchasing power parity dollars (PPP$). Only a few researchers employ GDP measured in market exchange rates which understates the income of poor countries. *Hill* and *Magnani* (2000) argue that we should use net domestic product defined as GDP minus depreciation. This makes sense, but gives similar results. The use of 'green GDP' which *Hill* and *Magnani* (2000) also propose is less advisable since we are interested in the relationship between pollution and economic activity which is best measured by GDP. Finally, *Perrings* and *Ansuategi* (2000) also estimate the PIR with consumption and with the human development index (HDI). However, the results for these indicators are mainly the same as for GDP and they even have lower explanatory power, i.e. lower adjusted R^2. Hence, the almost exclusive use of GDP in PPP$ is reasonable.

2.3 Traditional studies: Cross country analyses

2.3.1 Estimation technique

The traditional and still dominant technique in the EKC literature is to estimate

$$P_{it} = \alpha + \beta_1 Y_{it} + \beta_2 Y_{it}^2 + \beta_3 Y_{it}^3 + \beta_4 X_{it} + \epsilon_{it} \qquad (2.1)$$

where P is pollution measured in emissions per capita or in the concentration of a certain pollutant, Y is per capita GDP mostly measured in PPP$, i is an index for the country or for the measuring station, t is an index for time, X is a vector of additional explanatory variables, and ϵ is the error term. The vector X typically consists of the population density[3] and – for concentration data – of variables describing the site of the monitoring station (dummies for central city or suburban and for industrial, commercial, or residential land use nearby and the like). Sometimes (2.1) is also estimated without any additional explanatory variables, i.e. without $\beta_4 X_{it}$.

To have a reasonably large data set (2.1) is estimated using panel data from as many countries and years as possible. Some data sets are only available for one year so that a pure cross country analysis is applied. Panel data, however, are preferable as they allow for higher efficiency of the econometric estimates (*Ansuategi* 2000, 32). While many studies use ordinary least squares (OLS) to estimate (2.1), others employ generalized least squares (GLS) which is more appropriate as it corrects for heteroscedasticity and autocorrelation.

Some researchers estimate (2.1) without the cubic term $\beta_3 Y_{it}^3$. This technique, however, is too favourable to the EKC hypothesis. The cubic formulation, in contrast, allows for both an inverted U-shape and monotonically rising pollution. Additionally, it has the advantage that the EKC does not have to be symmetric as in the quadratic formulation: The cubic function might rise faster than it declines or vice versa. If the cubic term is not significant, it can be skipped.[4] This procedure – proposed by *Shafik* and *Bandyopadhyay* (1992) – has been followed by most researchers. If β_2 is also insignificant in the quadratic estimation, even the linear formulation – with both $\beta_2 Y_{it}^2$ and $\beta_3 Y_{it}^3$ dropped – makes sense. Therefore the estimation of (2.1) will produce one of the following seven results: A monotonically falling or rising PIR, an inverted U-shape (i.e. the EKC), a U-shape, an N-shape (first

[3]The population density can have a positive or negative sign and is often insignificant.

[4]It is possible that β_1, β_2, and β_3 are all insignificant in the cubic estimation, but that β_1 and β_2 are highly significant in the quadratic estimation (see table 1 in *Shafik* and *Bandyopadhyay* 1992).

rising, then falling, and finally rising again), an inverted N-shape or an insignificant, i.e flat PIR.

Estimations of (2.1) have not only been run with levels data but also with logarithms ($\ln P$ and $\ln Y$ instead of P and Y). Although most researchers estimate (2.1) in levels only or in logarithms only, it is actually an empirical question which functional form fits the data better. We should therefore estimate both versions – and also semi-logarithmic versions with only pollution or only income in logarithms – and pick the one that has the highest explanatory power. For a quadratic formulation the use of logarithms is theoretically preferable because as income goes to infinity, the estimation in levels predicts that pollution goes to minus infinity whereas when the logarithm of pollution goes to minus infinity, pollution approaches zero (*Cole et al.* 1997). This is an important theoretical point as emissions and concentrations cannot be negative. However, what really matters in practice is which functional form has the higher explanatory power inside the data range. This can be the estimation in levels or in logarithms (see table 4 in *Cole et al.* 1997). Furthermore, as long as the trough of the N-shape is outside the data range, a slowly declining PIR can also be estimated in levels. Note also that the turning point estimate depends on the choice of the functional form (*Hilton* and *Levinson* 1998). Hence, we should not have too much confidence in the turning point estimate – also because the data quality is poor (*Ekins* 1997).

Often (2.1) is estimated with a time trend, i.e. with an additional term $\beta_5 t$. This time trend is added in order not to attribute to GDP growth environmentally beneficial or harmful effects which go back to technological progress or to enhanced environmental awareness. However, as discussed in paragraph 2.3.3.3 below the introduction of a time trend is problematic.

Finally, (2.1) is often estimated using fixed effects, i.e. replacing the intercept α by a dummy variable α_i for each country or monitoring station. Hence, each country has its own intercept (see

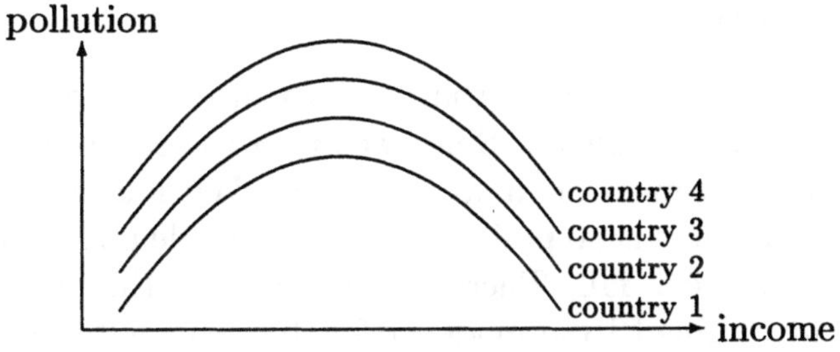

Figure 2.1: Estimation with fixed effects.

figure 2.1). This corrects for different climatic conditions (emissions by heating in the winter), different resource endowments (high stocks of fossil fuels or many possibilities for water power generation), and systematic measurement errors. Random effects are often calculated in addition to fixed effects. The assumption for random effects is that the α_i are drawn from a distribution with mean α. It is then checked with a Hausman test whether fixed or random effects are preferable (*Green* 1997, 632–634).

Note that (2.1) is a reduced form (*Panayotou* 1997). Actually, the influence of income on environmental regulation, on technology, and on structural change should first be estimated. Then the relationship between these variables and pollution could be examined. However, due to data problems income is used as a catch-all variable in (2.1). Thus we cannot tell which underlying cause is the main driving force behind the EKC: The estimation is a black box. Therefore, the results have given rise to widely varying interpretations. For instance, the results have been used to promote growth strategies as discussed in chapter 1 although, in fact, they are not suitable for policy recommendations.

2.3.2 Results

In this subsection we present the results which have been found using cross country analyses. Since we are considering man-made pollution, the PIR must begin at the origin: Without any income (or production) there is no pollution. So man-made pollution must first rise before it can fall. Therefore we interpret the finding of a monotonically falling PIR as evidence for the EKC, but the turning point of the EKC occurs at such a low income level that it could not be detected with the data at hand. Similarly, we interpret a U-shaped PIR as evidence of an N-shaped PIR. In this way of interpretation we follow part of the literature.[5] Furthermore, following *Cole et al.* (1997) and *Stern* and *Common* (2001), we consider the finding of an EKC with a turning point outside the sample range as evidence for a monotonically rising PIR.

Tables 2.1–2.5 summarize the estimation results. For each pollutant the first column shows our interpretation of the shape of the estimated PIR which is either monotonically rising (↗), an EKC (⌒), an N-shape (∼), or an insignificant relationship (is). In tables 2.1 and 2.2 results for emissions are indexed by an ^{e} while all other results represent concentrations. The second column gives either the income level at the turning point of the EKC (in thousands of mostly 1985 PPP$), the income range in which the turning points fall if several regressions have been run (for example in levels and logarithms, with and without fixed effects, or with and without additional regressors, i.e. X_{it} in (2.1)), the two turning points (peak and trough of the N-shape), or the actual estimation result (for example a monotonically falling PIR (↘) or a U-shape (⌣) and the turning point).

The main finding from tables 2.1–2.4 is that an EKC emerges for local and flow pollutants, whereas the PIR is monotonically rising for global and stock pollutants.

[5]See *Kaufmann et al.* (1998), *Scruggs* (1998), *Bradford et al.* (2000), *Dinda et al.* (2000), *Harbaugh et al.* (2000), and *Millimet* and *Stengos* (2000).

Table 2.1: Empirical results for the PIR for several air pollutants.

	SO_2		SPM				NO_x	
			dark matter		heavy particles			
Grossman and Krueger 1991	∼	4.5, 15.0	∼	5.0, 10.0	⌒	↘		
	⌒	4.5	∼	4.0, 10.0	↗	↗		
Shafik/Bandyopadhyay 1992	⌒	3.7		⌒	3.3			
Shukla and Parikh 1992[a]	is		is		⌒	11.0-12.5		
Grossman et al. 1994	⌒	13.9		↗	⌒ 27.5		⌒	19.4
Selden and Song 1994	⌒[e]	8.7-10.7		⌒[e]	9.5-10.3		⌒[e]	11.2-21.8
Grossman 1995	∼	13.4, 24.0		⌒	15.9		⌒	18.5
Grossman and Krueger 1995	∼	4.1, 13.5	⌒	6.2	⌒	↘		
Panayotou 1995	⌒[e]	2.9		⌒[e]	4.5		⌒[e]	5.5
Carson et al. 1997	⌒[e]	↘		⌒[e]	↘		⌒[e]	↘
Cole et al. 1997	⌒[e]	5.7-6.9		⌒[e]	7.3-8.1		⌒[e]	14.7-15.1
Lim 1997	⌒	↘		⌒	↘		⌒	↘
Panayotou 1997	∼	5.0, 17.5						
Vincent 1997				↗	↗			
Kaufmann et al. 1998	∼	⌣ 12.5						
Scruggs 1998	⌒	↘		⌒	↘			
Torras and Boyce 1998	∼	3.9, 15.4	∼	4.4, 10.5	⌒	↘		
Wu 1998			⌒	↘	⌒	n.a.		
Gallet et al. 1999[b]	⌒[e]	10.5					⌒[e]	14.5
List and Gallet 1999	⌒[e]	22.6					⌒[e]	8.7
Ansuategi 2000, chapter 5	⌒[e]	4.7-6.6						
Barrett and Graddy 2000[c]	∼	5.2, 12.5	⌒	8.1	⌒	↘		
	∼	1.8, 14.7	⌒	8.8	↗	↗		
Bradford et al. 2000	⌒	1.9-3.1	⌒	12.0	is			
Cavlovic et al. 2000	⌒	18.9	⌒	4.9	⌒	6.1	↗	⌒ 30.8
Cole 2000b, 112	⌒[e]	5.2					⌒[e]	20.3
Dinda et al. 2000	⌒	↘		∼	⌣ 12.5			
Hill and Magnani 2000	⌒[e]	7.4-12.7					⌒[e]	8.9-17.4
Hung and Shaw 2000	is			is			⌒	12.8
List and Gerking 2000	⌒[e]	26.1					⌒[e]	15.2
Millimet et al. 2000	⌒[e]	8.0					⌒[e]	12.0
Perrings and Ansuategi 2000	⌒[e]	9.6						
Gangadharan/Valenzuela2001	is			is			is	
Stern and Common 2001	⌒[e]	9.2						
	↗[e]	⌒ 908.2						

[a] Own calculations from table 3 in *Shukla* and *Parikh* (1992). [b] Own calculations from table 2 in *Gallet et al.* (1999). [c] Own calculations from tables 2 and 7 in *Barrett* and *Graddy* (2000). [e] Emissions, all other estimates represent concentrations.
⌒ = EKC. ↗ = Monotonically rising PIR. ↘ = Monotonically falling PIR. ∼ = N-shaped PIR. is = insignificant PIR. n.a. = Turning point estimate not available. numbers = Income level at turning points in thousands of mostly 1985 PPP$.
Notes: The first column for each pollutant shows our interpretation of the estimated shape of the PIR. The second column gives the turning point, the peak and the trough of the N-shape, the actual estimation result, or the range in which the turning points fall if several regressions have been run.

Table 2.2: Empirical results for the PIR for some further air pollutants and CO_2.

	CO		lead		VOC		CO_2 emissions	
Holtz-Eakin and Selden 1992							↗^{e}	⌒ 35.5-8000
Shafik/Bandyopadhyay 1992							↗^{e}	↗
Grossman et al. 1994	⌒	16.7	⌒	16.5				
Selden and Song 1994	is^e							
Grossman 1995	⌒	22.8	↗	↗				
Sengupta 1996							∼^{e}	8.7, 15.3
Carson et al. 1997	⌒^{e}	↘			⌒^{e}	↘	⌒^{e}	↘
Cole et al. 1997	⌒^{e}	9.9-10.1					↗^{e}	⌒ 25.1-62.7
Lim 1997							↗^{e}	↗
Moomaw and Unruh 1997							∼^{e}	12.8, 18.3
Roberts and Grimes 1997, 192							↗^{e}	↗
Hilton and Levinson 1998			⌒^{e}	4.0-11.0				
Schmalensee et al. 1998							⌒^{e}	9.8
Agras and Chapman 1999							↗^{e}	⌒ 101.0
							⌒^{e}	13.6
Galeotti and Lanza 1999a							⌒^{e}	17.1
Gallet et al. 1999[a]	⌒^{e}	13.8			⌒^{e}	15.0		
Borghesi 2000							↗^{e}	↗
Cavlovic et al. 2000	↗	⌒ 30.8					↗^{e}	⌒ 199.3
Hill and Magnani 2000							⌒^{e}	8.7-12.3
							↗^{e}	↗
Hung and Shaw 2000	⌒	6.8						
Perrings and Ansuategi 2000							↗^{e}	↗
Auffhammer et al. 2001b, 20							↗^{e}	⌒ n.a.
Azomahou and van Phu 2001							↗^{e}	↗
Bengochea-Morancho et al.2001							↗^{e}	↗
Carlsson and Lundström 2001							↗^{e}	↗
							⌒^{e}	24.6
Gangadharan/Valenzuela 2001							∼^{e}	5.8, 18.9
Heil and Selden 2001							↗^{e}	⌒ 36.0-2300

[a] Own calculations from table 2 in *Gallet et al.* (1999).

Note: For further explanations see table 2.1.

As can be seen from tables 2.1 and 2.2 most studies agree that there is an EKC for all air pollutants: SO_2, SPM, NO_x, CO, lead, and VOC. However, the turning point estimates vary widely. While some estimates lie at relatively low income levels, others are at the income level of the richest countries (or even beyond). So we cannot draw any definite conclusions concerning the income level at the turning point. The finding of an EKC for these pollutants is not surprising because these pollutants cause severe local health risks and because abatement is relatively cheap (*Selden* and *Song* 1994, 155). It seems that those pollutants causing the highest health risks – SO_2 and SPM (*Grossman* 1995) – are addressed at the lowest income levels. Although all these pollutants are stock pollutants, they all have very short lifetimes in the atmosphere and can therefore be considered as flow pollutants from a long-run point of view: the lifetime of SO_2 is 1–4 days,[6] that of NO_x is 2–5 days, that of CO is 1–3 months, and that of hydrocarbons (VOCs are hydrocarbons, see *Nentwig* 1995, 451) ranges from hours to a few days (*Liu* and *Lipták* 2000, 32 and *IPCC* 1996, 92, for a definition of lifetime see *IPCC* 1996, 76). SPM is washed out by rain- and snowfalls (*Liu* and *Lipták* 2000, 34) and thus has also a short lifetime.[7]

Some studies have found that air pollution starts to rise again at high income levels. For instance, *Kaufmann et al.* (1998, 216) find a U-shape for SO_2 with trough at 12500 PPP\$. They interpret this result as follows: At low income levels (below 3000 PPP\$, the lowest income level in their data set) SO_2 rises because of a shift from biomass fuels to coal. Then the SO_2 concentration falls as the economy moves away from fuels with high sulphur content such as coal towards fuels with low sulphur content such as oil and natural gas. Beyond 12500 PPP\$ the rising energy consumption and the increasing abatement costs outweigh the ongoing substitution of fuels and the SO_2 concentration rises again.

For CO_2 emissions, however, the majority of studies finds a monotonically rising PIR (see table 2.2). Nevertheless, some studies find an EKC, while others obtain that the slope of the PIR becomes increasingly steeper (*Shafik* and *Bandyopadhyay* 1992, *Borghesi*

[6]Since all empirical EKC studies consider SO_2 and NO_x only as air pollutants, we treat them as such. We therefore neglect that they are also stock pollutants causing acidification of soils, fens, and lakes.

[7]I could not find any data for the lifetime of lead.

2000, and *Perrings* and *Ansuategi* 2000). The PIR tends to be rising because CO_2 is a global stock pollutant with a lifetime of about 125 years (*Frey et al.* 1991, 165).

Note that even if we find an EKC for emissions per capita, pollution can continue to rise because of two reasons. First, the population is growing. Although population growth is one of the main driving forces behind environmental decay and stopping it is important (*Sengupta* 1996, *Wackernagel* and *Rees* 1997a, 168, and *de Bruyn* 2000, 11), we will not analyse this issue in this book.[8] Second, although emissions are falling, they might still lie above the assimilative capacity of nature such that concentrations continue to increase. However, since for short-lived air pollutants the EKC is also found for concentrations (at least in urban areas), this does not seem to constitute a problem. But for the long-lived CO_2 the situation is different (cf. subsections 3.10.1 and 3.10.2).

For the river water quality indicators in table 2.3 the evidence is weaker. While many EKCs (with widely varying turning point estimates) are found, several insignificant PIRs are also obtained. Again some results indicate an N-shape. *Shafik* and *Bandyopadhyay* (1992) explain the final upturn of the N-shape for fecal coliform with the fact that people in high income countries no longer directly depend on river water quality because of improvements in water supply systems. Since the monitoring stations are in rivers, the concentrations of these pollutants would quickly decline if emissions stopped: So river pollutants are short-lived. Thus they can also be viewed as flow pollutants.

Lack of clean drinking water and lack of urban sanitation are classic examples of strictly local problems. Furthermore, the health risks are high (survival is at stake) and the costs of provision are fairly low (*Shafik* 1994, 761). Hence, it is not surprising that the EKC has a very low turning point (see table 2.4): As soon as resources become available, these two problems are solved. However,

[8]For a study of the relationship between the PIR and population growth see *Baldwin* (1995).

Table 2.3: Empirical results for the PIR for the concentration of river water pollutants.

	oxygen regime			pathogens			heavy metals		
	pollu-tant			pollu-tant			pollu-tant		
Shafik and Bandyopadhyay 1992	−DO[a]	is		FC	∼	1.4, 11.4			
Grossman and Krueger 1995	−DO[a]	⌒	2.7	FC	⌒	8.0	lead	∼	1.9, 14.5
	BOD	is		TC	∼	3.0, 8.5	cadmium	⌒	11.6
	COD	is					arsenic	∼	4.9, 15.0
	nitrates	⌒	10.5				mercury	is	
							nickel	is	
Cole et al. 1997	nitrates	⌒	15.6-25.0						
Lim 1997	BOD	⌒	n.a.						
Vincent 1997	BOD	is							
	nitrogen	↗	↗						
Scruggs 1998	−DO[a]	⌒	↘	FC	↗	↗			
Torras and Boyce 1998	−DO[a]	⌒	↘	FC	is				
Barrett and Graddy 2000[b]	−DO[a]	⌒	3.0	FC	⌒	9.5	lead	is	
	BOD	is		TC	∼	3.1, 8.9	cadmium	⌒	11.1
	COD	is					arsenic	↗	↗
	nitrates	⌒	10.5				mercury	is	
							nickel	∼	⌣ 9.9
Bradford et al. 2000	−DO[a]	⌒	1.2	FC	↗	↗	lead	∼	⌣ 19.8
	BOD	⌒	3.2	TC	∼	⌣ 9.0	cadmium	is	
	COD	⌒	↘				arsenic	⌒	13.2
	nitrates	∼	⌣ 10.6				mercury	is	
							nickel	∼	⌣ 12.6
Cavlovic et al.2000	aggregate	⌒	6.5				aggregate	⌒	16.4
Hettige et al. 2000	BOD	⌒[c]	7.0						

[a] Since more dissolved oxygen means less pollution, the results are for minus dissolved oxygen. [b] Own calculations from table 5 in *Barrett* and *Graddy* (2000). [c] This PIR rises until 7.0 and then stays constant.
DO = dissolved oxygen. BOD = biological oxygen demand. COD = chemical oxygen demand. FC = fecal coliform. TC = total coliform.
Note: For further explanations see table 2.1.

Ekins (1997, 811) argues that these two indicators are measures of the lack of basic services, not of environmental quality.

For deforestation the evidence is weakest (see table 2.4). While some studies find an insignificant PIR, others find an EKC, but

Table 2.4: Empirical results for the PIR for some further pollutants.

	lack of clean water		lack of sanitation		deforestation		municipal waste	
Shafik and Bandyopadhyay 1992	$\frown$	$\searrow$	$\frown$	$\searrow$	is		$\nearrow$	$\nearrow$
Cropper and Griffiths 1994					$\frown$	4.8-5.4		
					is			
Panayotou 1995					$\frown$	0.8		
					is			
Lugo 1996					$\frown$	n.a.		
Cole et al. 1997							$\nearrow$	$\nearrow$
Lim 1997					$\nearrow$	$\nearrow$	$\frown^a$	n.a.
							$\nearrow^b$	$\nearrow$
Torras and Boyce 1998	$\frown$	$\searrow^c$	$\frown$	$\searrow^c$				
Cavlovic et al. 2000					$\frown$	4.9		
Perrings and Ansuategi 2000	$\frown$	$\searrow$			$\frown$	1.9		
Gangadharan and Valenzuela 2001					is			
Minliang et al. 2001							$\nearrow$	$\nearrow$

[a] Estimate for domestic waste. [b] Estimate for industrial waste. [c] Actually an inverted N-shape with rising pollution in the range of 11.3-14.9 or 11.0-16.9, respectively.

Note: For further explanations see table 2.1.

the explanatory power is very low ($R^2 = 0.09$ in *Perrings* and *Ansuategi* 2000). So it seems that income is not an important determinant of deforestation. However, these weak results might also reflect the poor data quality discussed in section 2.2.

The PIR for municipal waste, finally, is ever rising (see table 2.4). Municipal waste is a stock pollutant since it is deposited and accumulates in waste disposal sites.[9] Furthermore, these disposal sites can be located at places where nobody lives or only poor people who have no political power (*Shafik* 1994). Thus since the waste problem can easily be externalized, waste generation is not reduced.

[9] If waste is incinerated, about 30% of its original weight remain for disposal (*Nentwig* 1995, 347).

The main focus of all empirical studies is the income level at the turning point while the pollution level at the turning point is not even reported in most studies. What really matters, however, is the pollution level since it determines what damages occur and whether or not thresholds are overstepped (*Agras* and *Chapman* 1999, *Borghesi* 1999, and *Perrings* and *Ansuategi* 2000). We will analyse thresholds in section 3.10.

2.3.2.1 Results for aggregate measures of pollution

It has been argued that pollution as perceived by the general public is an aggregate measure (*Wu* 1998). Hence, to obtain an overall picture of the PIR, there have also been efforts to estimate the PIR for aggregate pollution. However, aggregation of different pollutants is 'problematic' (*Magnani* 2000a, 433). The weights of different pollutants in the aggregate are always somewhat arbitrary. Thus all aggregates can only be considered as approximations (*de Bruyn* 2000, 67–70). Some researchers (*Ekins* 1997, *Scruggs* 1998, and *Wu* 1998) aggregate the pollutants for which they have data using simple weighting techniques. Not surprisingly, the results mirror the average of the PIR estimates for the single pollutants. Therefore we concentrate on aggregates which use different data sources than the studies analysed so far. The results for these aggregate indicators are shown in table 2.5.

The most frequently analysed aggregate indicator is energy consumption – the 'chief source of a number of environmental problems' (*Suri* and *Chapman* 1998, 196). The traffic volume can also be considered as the root of several types of pollution. Both indicators are ever rising with income. *Hettige et al.* (1992) find evidence that emissions of 320 toxic substances per unit of industrial output increase with income. This implies that the PIR for toxic substances is also monotonically rising because industrial output increases with income (*Ekins* 1997, 812). In contrast, *Carson et al.* (1997) obtain an EKC for air toxics emissions while *Millimet* and *Stengos* (2000) estimate an N-shaped PIR for

Table 2.5: Empirical results for the PIR for some aggregate measures of pollution.

Cole et al. 1997	energy consumption	↗	⌒ 22.5-34.7
Suri and Chapman 1998		↗	⌒ 55.5
Agras and Chapman 1999		↗	⌒ 62.0-195.0
Cole et al. 1997	traffic volume	↗	⌒ 65.3-108.2
Hettige et al. 1992	air toxics emissions	↗	↗
Carson et al. 1997		⌒	↘
Millimet and Stengos 2000		∼	⌣ 30.0
Wang et al. 1998	health risk of hazardous waste sites	⌒	22.6
De Bruyn and Opschoor 1997	throughput	∼	n.a.
Rothman 1998	ecological footprint	↗	↗

Note: For explanations see table 2.1.

650 toxic chemicals. Analysing subcategories of these chemicals *Millimet* and *Stengos* (2000) also find weak evidence that subcategories with mainly local effects have turning points at smaller income levels than subcategories with more transboundary effects. For the health risk of hazardous waste sites *Wang et al.* (1998) derive an EKC with a turning point at high income levels. *De Bruyn* and *Opschoor* (1997) construct a measure of the flow of material and energy. This throughput indicator is aggregated from the consumption of steel, energy, and cement and from weight of freight transport. The estimated form is an N-shape, but the turning points have not been calculated. The ecological footprint – developed by *Wackernagel* and *Rees* (1997a) – is a related, but more detailed indicator. It converts material and energy consumption flows into the area of land and water which is necessary to make these flows sustainable.[10] *Rothman* (1998) estimates a monotonically rising PIR for the ecological footprint per capita.

[10]'The ecological footprint [of an economy] can be represented as the aggregate area of land and water [...] that is claimed by participants in that economy to produce all the resources they consume, and to absorb all their wastes they generate on a continuous basis' (*Wackernagel* and *Rees* 1997b, 7). The area of land is an aggregate of various ecological categories: Built-up area,

Moreover, *Nentwig* (1995, 547–551) observes that the list of endangered species – a crude measure of biodiversity loss – is growing steadily and that the anthropogenic speed with which species become extinct is ten thousand times higher than the natural one resulting in huge losses in biodiversity. Similarly, *Skonhoft* and *Solem* (2001) find that wilderness land – a crude measure of biodiversity – is declining steadily with income. Finally, *Ekins* (1997) summarizes two reports of the OECD and of the European Commission on the state of the nature. These reports acknowledge that progress has been made with some urgent pollution problems – those problems for which an EKC has been found. However, there remain substantial problems 'across all areas of environmental concern' (*Ekins* 1997, 815). Furthermore, high-risk activities are increasing. In fact, OECD countries will 'have to face more intractable problems than those solved in previous decades' (*Ekins* 1997, 815). Thus environmental degradation grows despite heavy investment in abatement. These reports therefore clearly negate the existence of an EKC for overall pollution.

Summarizing we find that the EKC has been observed for pollutants with mainly local and immediate effects such as short-lived air pollutants. It seems that those air pollutants causing the highest health risks are addressed at the lowest income levels. For water pollutants – which can also be considered as flow pollutants – the EKC has also been found although the evidence is less clear. For global and stock pollutants such as CO_2, however, the PIR for emissions seems to be monotonically rising.[11] Furthermore,

arable land, pasture, forest, and land used to absorb CO_2 emissions caused by burning fossil fuels (*Wackernagel et al.* 1999, 381).

[11] *Cavlovic et al.* (2000) draw a similar conclusion from a meta-analysis of EKC studies. A meta-analysis is a method to statistically summarize the findings of different studies with conflicting results. However, *Cavlovic et al.* (2000) do not distinguish between estimations in levels and in logarithms (cf. subsection 2.3.1) and between estimations with and without a time trend (cf. paragraph 2.3.3.3).

most aggregate pollution indicators are ever rising. Hence, EKC studies have observed improvements in well-known problems, but they have missed more recent concerns (*Ekins* 1997, 816).

Many studies obtain N-shaped PIRs. Although some authors have argued that the final upturns are not important (*Grossman* and *Krueger* 1991 and 1995) or that they are merely a 'curiosity' (*Ansuategi* 2000, 41), they are not at all irrelevant (*Torras* and *Boyce* 1998, 157): The final upturns mean that in the end income growth is not sustainable. However, in the studies to date there is surprisingly little interest in the final upturn of the N-shaped PIR (*Stern* 1998, 192).

2.3.3 Critique of the traditional estimation technique

The traditional estimation technique with which these results have been derived is often criticized. Estimations with (2.1) might suffer from the following weak points which we will discuss in turn: A simultaneity bias might hamper the results, other functional forms than the polynomial might fit the data better, the use of a time trend might be problematic, the regression might suffer from multicollinearity, income might not have an immediate, but only a lagged effect on pollution, and a homogeneity test might reveal that the slope coefficients are not identical in all countries. Additional objections against the traditional estimation technique will be discussed in section 2.4.

2.3.3.1 Simultaneity bias

Income does not only influence pollution, but pollution has also an effect on income: Labour productivity is reduced because some pollutants decrease the ability to concentrate and learn and because other pollutants cause work day losses due to health problems (*van Ewijk* and *van Wijnbergen* 1995, 197). Moreover, pollution reduces harvests, forestry yields, and fish catches (*McConnell*

1997, 392). If pollution has an effect on income, there is a simultaneity bias in the regression and OLS produces biased and inconsistent results (*Stern et al.* 1996). However, only *Hung* and *Shaw* (2000) find simultaneity in two of their four regressions, while no simultaneity is detected by *Holtz-Eakin* and *Selden* (1992), *Cole et al.* (1997), *List* and *Gallet* (1999), and *List* and *Gerking* (2000). When *Hung* and *Shaw* (2000) estimate a simultaneity model, they obtain similar results than with the traditional model. Hence, although pollution surely influences income, this effect is mostly so small that it is not significant when tested for. Thus it seems that the results do not suffer from a simultaneity bias.

2.3.3.2 Other functional forms

While most studies use a quadratic or cubic polynomial to estimate the PIR, some studies try to find better functional forms. In fact, in the first EKC study *Grossman* and *Krueger* (1991, 12) first use dummy variables for different income ranges. This estimation shows that the cubic function 'fits the data fairly well'. *Schmalensee et al.* (1998) use a spline function which is piecewise linear: They split up the data in ten income groups of equal size and estimate a linear function for each group where these functions are restricted such that the ensuing PIR is continuous. A spline function with only four income groups is estimated by *Hilton* and *Levinson* (1998). Both studies derive similar results with the spline function and with the traditional polynomial.

Galeotti and *Lanza* (1999a) propose to use the functional form of the Gamma or Weibull distribution. These two distributions can be monotonically falling or inverted U-shaped where the inverted U-shape is more or less asymmetric (rising fast and falling slowly). But they cannot be monotonically rising or N-shaped. Thus with this specification we surely find falling pollution at high income levels. This biases the results in favour of the EKC hypothesis. Although *Galeotti* and *Lanza* (1999a, 1) state that Gamma and Weibull are unambiguously better than the traditional technique, their table 7 reveals that this is not always the case and that the differences are small. Since the Gamma and Weibull distribution favour the EKC hypothesis too much, it seems preferable to use the traditional functional form.

Bradford et al. (2000) believe that the income growth rate is also important for the EKC. Therefore they enter not only income levels into the regression function, but also income growth rates. However, as tables 2.1 and 2.3 reveal, their results are similar to those of other studies employing the traditional regression function (2.1).

It seems clear that in reality the PIR of a country is not exactly a cubic function. This is only an approximation to the actual path. Hence, when allowing for a very flexible functional form it is not surprising that *Millimet* and *Stengos* (2000), *Millimet et al.* (2000), and *Azomahou* and *van Phu* (2001) find that this gives significantly better results than the traditional polynomial. These authors use nonparametric approaches in which the value of pollution P_0 for a given income level Y_0 is given by[12]

$$P_0 = \sum_{i=1}^{n} \frac{K[(Y_i - Y_0)/s]}{\sum_{j=1}^{n} K[(Y_j - Y_0)/s]} P_i$$

where $K[\cdot]$ is the standard normal density function, s is a smoothing parameter, and n is the sample size. Hence, P_0 is a weighted average of all observed P_i where the weight is smaller when the corresponding income level Y_i is farther away from Y_0. In most cases the results from nonparametric regressions are similar to those from the traditional polynomial although there are exceptions.[13] While *Millimet* and *Stengos* (2000) state that the polynomial tends to overestimate the income level at the turning point, *Millimet et al.* (2000) derive the opposite result. Hence, the approximation by

[12]This equation is taken from *Azomahou* and *van Phu* (2001, 8) and corrected according to *Härdle* (1990, 25).

[13]The differences in *Azomahou* and *van Phu* (2001), however, are probably due to outliers in the data.

a polynomial is often sufficiently exact to show the general shape of the PIR. Moreover, although the nonparametric approach allows significant improvements, it can also be 'overly computationally burdensome' (*Millimet* 2000, 5).

2.3.3.3 Time trend

The time trend is introduced into the estimation to account for changes in technology and in environmental awareness which are not related to income. However, as *de Bruyn et al.* (1998, 165–167) show, the use of a time trend has important effects: Suppose that we estimate (2.1) including a time trend and that the time trend is negative. Then the estimated EKC with a turning point at Y_3 shifts downwards over time as shown in figure 2.2a. However, a single country which is growing over time follows the dashed path in figure 2.2a: The country does not move along the estimated EKC. Hence, the turning point for this country will occur at the smaller income level Y_2. Even worse, if the time trend is insignificant, the EKC might first shift downwards and then upwards as in figure 2.2b. Then the country follows an N-shaped path although an EKC has been estimated. If the time trend is positive, we might even estimate an EKC although the true PIR is monotonically increasing. Hence, *to conclude that pollution is falling at high income levels it is crucial that the time trend is significantly negative.* However, as the estimation results reveal, the time trend is not always negative, but it can also be insignificant or even significantly positive. Thus the use of a time trend calls the EKC results into question. So it might be advisable to reestimate the PIR without the time trend.

Cole (2000b, 79–85) argues that whenever a time trend is used in the estimation, the average income growth rate and the estimated time trend should be used to calculate the true income turning point (Y_2 in figure 2.2a). Except for *Cole* (2000b), however, all researchers using a time trend calculate the turning point neglecting the time trend, i.e. they derive Y_3 in figure 2.2a. It turns

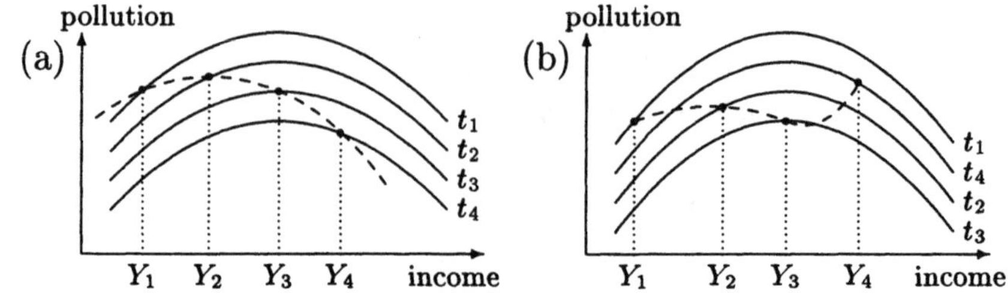

Figure 2.2: The estimated EKC and the PIR of a growing country over time when the time trend is (a) negative or (b) insignificant. (Source: De Bruyn et al. 1998)

out that Y_2 is often similar to the turning point estimate when the time trend is omitted, but that Y_2 and Y_3 can be clearly different (*Cole* 2000b, table 5.5). Hence, the turning point estimates of many studies are biased upwards if the time trend is negative or biased downwards if it is positive.

2.3.3.4 Multicollinearity

The problem of multicollinearity is completely disregarded by almost all EKC studies and the very few who calculate test statistics do not discuss the implications. In my opinion, however, multicollinearity should be a major concern. When the explanatory variables are highly correlated, small changes in the data can have 'dramatic' effects on the results (*Green* 1997, 420). We should worry about this because pollution data are unreliable. Thus – if multicollinearity is present – even slight measurement errors can cause wide swings in the estimated turning point and might even change the estimated shape of the PIR. Furthermore, since GDP, squared GDP, and cubed GDP enter the regression function, there is reason to believe that multicollinearity might actually be present. In fact, *Stern* (1999, 13) finds evidence for multicollinearity in three of his six estimations although he only uses a quadratic function. In contrast, in *Ansuategi's* (2000, 94) regression multicollinearity does not constitute a problem.

Grossman and *Krueger* (1995) have made the following estimation technique popular:

$$P_{it} = \alpha + \beta_1 Y_{it} + \beta_2 Y_{it}^2 + \beta_3 Y_{it}^3 + \beta_4 \bar{Y}_{it-} + \beta_5 \bar{Y}_{it-}^2 + \beta_6 \bar{Y}_{it-}^3 + \beta_7 X_{it} + \epsilon_{it} \tag{2.2}$$

where $\bar{Y}_{it-}$ is the average of GDP per capita over the prior three years and where the other variables are defined as in (2.1). *Grossman* and *Krueger* (1995, 361) include $\bar{Y}_{it-}$ to capture the effect of permanent income and because past income levels influence current environmental standards. This estimation technique has subsequently been used by several researchers (*Grossman et al.* 1994, *Islam* 1997, *Barrett* and *Graddy* 2000, and *Harbaugh et al.* 2000). Since current GDP, Y_{it}, and lagged GDP, $\bar{Y}_{it-}$, are 'highly correlated' (*Grossman* and *Krueger* 1995, 361) and since both are used as cubic functions, it seems highly probable that multicollinearity is present. This is in fact the case: Most β_j, $j = 1, \ldots, 6$, are not significant, but jointly they are highly significant.[14] Furthermore, the turning point for dark matter in *Grossman* and *Krueger* (1995) is 6151 PPP\$, but only by dropping one country (Hong Kong) from the sample *Barrett* and *Graddy* (2000) obtain a turning point of 7286 PPP\$ – which lies more than two standard deviations away from 6151 PPP\$. These are both clear signs of multicollinearity (*Green* 1997, 420). Hence, the main result of *Harbaugh et al.* (2000, 2) – namely that the turning points and indeed the shape of the PIR are both 'sensitive to slight variations in the data and to reasonable permutations in the econometric specification' – does not invalidate the EKC hypothesis as they interpret it, but only clearly shows that (2.2) is hampered by severe multicollinearity problems. *Islam* (1997) derives the same

[14] See *Grossman* and *Krueger* (1995, 373–376), *Grossman et al.* (1994, tables 2a–7b), *Barrett* and *Graddy* (2000, tables 2–7), and *Harbaugh et al.* (2000, table 3 and appendix table 1–2). *Islam* (1997) does not show his tables.

main result – that the PIR differs widely – and also misses the most plausible explanation: multicollinearity.

Hence, when estimating the PIR, we should not use (2.2) and we should check whether or not multicollinearity is present. Furthermore, since *Stern* (1999) finds multicollinearity in his data although he uses only a quadratic specification, multicollinearity might even explain the large differences in the estimated turning points of different studies.

2.3.3.5 Lagged effects

Most researchers argue that environmental policy is a major driving force behind the EKC (see chapter 3). When income rises, environmental regulations become stricter. However, this process needs some time. First, the regulations are worked out and debated over in the parliament. Then it takes some time before the effects of the new regulations become measurable: For example, firms need time to adjust their technology (*Auffhammer et al.* 2001b). Furthermore, there are often temporary arrangements before the new regulations come into force. Nevertheless, the EKC is always estimated with current GDP. Only *Grossman* and *Krueger* (1995) and those using (2.2) have introduced lagged GDP. Actually, lagged GDP has a higher explanatory power than current GDP (*Grossman* and *Krueger* 1995, 366). It might be expected that the lag is longer than three years. Hence, *Harbaugh et al.* (2000) also estimated with the average GDP of the last ten years. Since, these estimations are all hampered by multicollinearity, it might be interesting to estimate the EKC with a polynomial of only lagged GDP and to see whether or not the explanatory power is really higher than with current GDP. However, to my knowledge this has not yet been done.

Another technique to account for lagged effects is to include lagged pollution, P_{it-1}, as an additional regressor in (2.1). Then an increase in current income has a direct effect on current pollution and through the lagged pollution regressor also on pollution in the next period and all future periods. This technique – used by *Agras* and *Chapman* (1999) and by *Auffhammer et al.* (2001b) – assumes that the largest effect of current income is on current, not future pollution levels. In contrast, this is not the case when using lagged GDP regressors. Moreover, a regression with a lagged pollution regressor might be hampered by unit roots as explained in section 2.4 below.

2.3.3.6 Homogeneity test

It has been criticized that PIR-estimations assume that all countries follow the same path. While an estimation with fixed or random effects allows for different levels of pollution, it is still assumed that the slope coefficients are identical, i.e that all countries reach the turning point at the same income level (see figure 2.1 on page 33). Instead of just imposing such a strong restriction, we should check with a homogeneity test whether or not this assumption is justified. If it is not, the estimates of cross country studies are biased and inconsistent (*List* and *Gallet* 1999, 411). Several researchers have carried out such a test (*Koop* and *Tole* 1999, *List* and *Gallet* 1999, and *de Bruyn* 2000, 105–106). They all derive highly significant evidence that the PIRs do not have the same slope coefficients in different countries. Using different techniques the same result is also found by several other authors.[15] Hence, *every country has its own PIR.* As the results of several studies[16] show these PIRs do not only have different turning points, but also exhibit different shapes. This explains the vast variation in results that cross country studies find: Depending on which countries are in the sample, the estimated PIR will be different.

[15]See *Dijkgraaf* and *Vollebergh* (1998), *Stern et al.* (1998), *Perman* and *Stern* (1999), *Millimet* (2000), *Millimet et al.* (2000), and *Auffhammer et al.* (2001b).

[16]See *Dijkgraaf* and *Vollebergh* (1998), *Stern et al.* (1998), *Koop* and *Tole* (1999), *List* and *Gallet* (1999), *Millimet* (2000), and *Millimet et al.* (2000).

It also follows that developing countries do not have to grow along the path that developed nations have taken as *Lucas et al.* (1992) have already supposed. The same result is also derived with a different technique by *Roberts* and *Grimes* (1997), *Hilton* and *Levinson* (1998), *Gallet et al.* (1999), and *List* and *Gerking* (2000): They show that the cross country PIR changes over time. Hence, a country starting its development later on follows a different PIR.

However, when estimating the PIR of each country separately, we actually consider a time series model. Hence, new problems arise which we discuss in the next section.

2.4 Time series analysis

One of the main shortcomings of time series EKC studies is that the time series are only short. Environmental data are simply not available over long time periods. Only for emission data relatively long time series are existing because emissions are estimated from fuel use data which had already been gathered before environmental data collection started. The lack of long time series is the main reason why most EKC studies use cross country analysis. However, as the homogeneity tests revealed, different countries have different PIRs. Hence, pooling is not allowed. The analysis of time series has the additional advantage that data within a country are better comparable as they are measured identically.

However, the possible presence of unit roots makes the statistical analysis more involved. A time series x is not stationary if it contains a unit root, i.e. if $\alpha = 1$ in

$$x_t = \alpha x_{t-1} + \epsilon_t$$

where ϵ is the error term. When x is not stationary, shocks (represented by ϵ) have permanent effects and do not evaporate over time (*Green* 1997, 847). Including a non-stationary variable in a regression implies that standard t-tests are invalid. Hence, it can lead us to conclude that there is a significant relationship although in fact there is none. 'Pretesting variables for unit roots is therefore a necessary routine to prevent the occurrence of spurious results' (*de Bruyn* 2000, 107). In fact, *de Bruyn* (2000) finds that GDP and emissions of SO_2, NO_x, and CO_2 in 4 OECD countries all contain unit roots except for SO_2 in the United States. *Liski* and *Toppinen* (2001) obtain unit roots for CO_2 emissions in 10 of 17 OECD countries. Moreover, *Perman* and *Stern* (1999) derive that SO_2 emissions and GDP have a unit root in almost all of 74 countries. When using more sophisticated panel unit root tests *Perman* and *Stern* (1999) find the same result, whereas *Heil* and *Selden* (2001), in contrast, do not find unit roots in CO_2 emissions and GDP. So the evidence is mixed.[17] Thus more research is needed to determine whether or not unit roots are present. Without unit roots estimation results from (2.1) – for each country separately – are valid. But if there are unit roots, the results may be spurious.

If all variables contain unit roots, there are two possible ways out of this problem. First, there might be cointegration: Pollution and income might move together such that shocks only cause deviations from a long-run equilibrium and have no permanent effects. This long-run equilibrium can be estimated with (2.1). If there is cointegration, the residuals of (2.1) are stationary. When testing the residuals of (2.1) for unit roots, *de Bruyn* (2000, 115) finds that only CO_2 in the Netherlands and in West Germany are cointegrated with income, but the PIR is monotonically rising. For the ten other estimations, however, there is no cointegration. Then shocks have permanent effects on the PIR.[18] Hence, *de Bruyn* (2000) finds no evidence for the EKC. *Perman* and *Stern* (1999,

[17] There is a 'considerable debate' in the macroeconomic literature whether or not GDP contains a unit root (see *de Bruyn* 2000, 109–110 and the citations therein).

[18] A possible shock which shifts the PIR upwards is described by *Setzer* (1998): In an economic recession energy consumption rises, but does not fall back to the initial level once growth accelerates again such that the PIR is path dependent.

34) find evidence for cointegration between SO_2 emissions and income in only 7–39 of 74 cases (for different estimation methods). However, individual country cointegration tests suffer from 'very low power' due to the moderate length of the time series (*Perman* and *Stern* 1999, 17). Hence, the probability is high that cointegration is rejected although it is present. Several different panel cointegration tests give mixed results, but *Perman* and *Stern* (1999, 23) conclude that there is weak evidence for cointegration in the panel as a whole. They also derive that each country has its own cointegration vector, i.e. its own PIR. This implies that the traditional technique might be valid as long as we allow the PIR to differ between countries. However, if there is no cointegration, previous results are spurious. More research is needed to find out whether or not there is cointegration.

The second possibility when both variables contain unit roots is to estimate the PIR in growth rates which are stationary. So we estimate

$$\Delta P_{it} = \alpha + \beta \Delta Y_{it} + \epsilon_{it} \tag{2.3}$$

where $\Delta x_{it} = lnx_{it} - lnx_{it-1}$ and where we have removed the quadratic term ΔY_{it}^2 because it is almost perfectly collinear with ΔY_{it} (*de Bruyn* 2000, 111). The results of *de Bruyn* (2000, 112) show that economic growth increases emissions of SO_2, NO_x, and CO_2 as β is significantly positive in most cases. So growth is bad for the environment. Without growth, however, pollution would decline as α is negative. Trying to find out why pollution declines *de Bruyn* (2000, 122–133) reestimates (2.3) with an additional regressor $\beta_2 \ln Y_{t-1}$. He finds that β_2 is always negative and mostly significant. Therefore economic growth directly increases pollution, but it also increases income and thus decreases pollution in all subsequent periods. However, only after 53–381 years are the initial costs of economic growth compensated by long-term benefits (*de Bruyn* 2000, 128). Hence, reductions in pollution occur not because of growth, but despite of growth. If there is cointegration, however, (2.3) gives only the short-run dynamics whereas (2.1) gives the long-run relationship (*Green* 1997, 852).

2.5 Emission forecasts

Some studies predict the future path of global emissions. First, population and GDP forecasts are taken from the existing literature or estimated. These are used to predict emissions in each country based on EKC results derived with the traditional technique. Then emissions are aggregated to world emission levels. All studies conclude that world emissions of SO_2, SPM, NO_x, CO, and CO_2 are rising fast until the endpoint of the forecast which is 2020, 2025, 2050, or 2100.[19] One of the main driving forces behind this result is that the income distribution is skewed towards zero (*Selden* and *Song* 1994, 156): Most people's income is below the world mean income level. If we considered only world mean income, SO_2 emissions per capita would decline from 55 to 38 kg between 1990 and 2025, and total emissions would increase by ten percent due to population growth. If we take the actual skewed income distribution into account, however, emissions per capita double (from 73 to 142 kg) while total emissions triple (*Stern et al.* 1996, 1157). Hence, emissions rise sharply although *Stern et al.* (1996) use *Panayotou's* (1995) SO_2 result with a turning point at 2900 \$ – the third lowest turning point of all studies (see table 2.1). An additional factor causing emissions to grow is that it is assumed that income and population are growing faster in poor countries than in rich ones. So CO_2 emissions rise with an average growth rate of 1.8% until 2100 (*Holtz-Eakin* and *Selden* 1992) and thus triple or even quadruple until 2100 (*Heil* and *Selden* 2001). CO_2 emissions

[19] For CO_2 see *Holtz-Eakin* and *Selden* (1992), *Schmalensee et al.* (1998), *Galeotti* and *Lanza* (1999b), *Auffhammer et al.* (2001b, only for China), and *Heil* and *Selden* (2001). For SO_2 and NO_x see *Selden* and *Song* (1994), *Stern et al.* (1996, only SO_2), and *Cole* (1999). For SPM and CO see *Selden* and *Song* (1994).

also grow fast because the EKC rises steeply but falls only slowly (*Galeotti* and *Lanza* 1999b).

However, such forecasts are not unproblematic. If there are unit roots without cointegration, shocks have permanent effects and forecasts are not feasible (*Labson* and *Crompton* 1993). So technological breakthroughs or new regulations might radically change the PIR we have observed in the data of the past. In fact, there is evidence that the PIR changes over time (*Roberts* and *Grimes* 1997, *Hilton* and *Levinson* 1998, *Gallet et al.* 1999, and *List* and *Gerking* 2000). Furthermore, we found that every country has its own PIR. Hence, forecasts based on cross country evidence are 'unwise' – especially when cross country evidence from OECD countries is used to predict the emissions of developing countries (*Stern* and *Common* 2001, 175). These shortcomings show that emission forecasts should not be taken too literally. However, these predictions caution against too optimistic conclusions: Even if there is an EKC with a low turning point (for example 2900 $), emissions are likely to increase for many years to come. Hence, if we really want to reduce emissions, drastic measures must be taken which go far beyond what has been done so far to combat pollution or we must hope for a technological revolution (*Holtz-Eakin* and *Selden* 1992).

2.6 Conclusion

Environmental problems differ in several aspects: The effects on human health, the availability and costs of abatement technologies, and the spatial and time distribution of costs and benefits of abatement. So the 'possibilities of 'growing out of environmental degradation' differ among problems' (*Ansuategi* 2000, 128). Hence, reviewing the empirical evidence for the EKC we find that the EKC is only observed for flow and local pollutants, but not for stock and global pollutants. Although the EKC is not found for all pollutants, the EKC has been interpreted to mean that pollution is only a temporary problem in the course of economic growth. However, even for the pollutants for which an EKC is observed this is a 'dangerous' conclusion because the forecasts show that 'temporary may entail a time scale of 100 years or more' and because 'certain types of environmental damage are irreversible' (*Hill* and *Magnani* 2000, 17). Irreversible damages will be analysed in section 3.10 below. Moreover, sometimes an N-shape is obtained. The final upturn means that in the end income growth is not sustainable. Nevertheless, most researchers do not give weight to this finding.

Furthermore, the EKC is estimated for a few pollutants only. For many other types of pollution, however, data are not available. Data tend to be collected for human health related indicators only. Besides, most aggregate measures of pollution show a monotonically rising PIR. Therefore growth does not improve the overall environmental quality. Hence, the conclusion that growth is necessary (*Asafu-Adjaye* 1998, 69) – or even sufficient (*Azomahou* and *van Phu* 2001, 3) – for falling pollution is not tenable or in the opinion of *Ayres* (1995, 97) it is 'false and pernicious nonsense'.

The empirical results show that the turning points and even the shape of the EKC vary widely between studies. This has led several researchers to relatively pessimistic conclusions concerning the EKC hypothesis: 'Empirical evidence is weak' (*Stern* 1998, 174). 'Unequivocal evidence for an EKC relationship is very scant' (*Ekins* 1997, 805). 'The EKC is only an empirical regularity that at best has proved to be a rule with too many exceptions' (*Ansuategi* 2000, 24). 'Whether economic growth will be beneficial or harmful to the environment in the long run remains a matter of belief' (*Neumayer* 1998, 172). 'The EKC hypothesis can be deemed invalid' (*Ekins* 1997, 813).

However, the vast variety of results can most probably be explained by weak points in the econometric analysis of the PIR such that the above conclusions are based on unsound grounds. First, the inclusion of a time trend in the regression, but the neglect of this time trend when calculating the turning point can bias the turning point estimate. Second, if there is multicollinearity, slight variations in the data can have dramatic effects on the results. Since the data sources are unreliable and since the regressors are correlated, multicollinearity might account for the vast variation in results. Third, homogeneity tests reveal that different countries follow different PIRs. Hence, depending on which countries are in a cross country analysis the turning point of the EKC or even the shape of the PIR is different. We therefore have to conduct a time series analysis of each country separately. This is unproblematic in the absence of unit roots. However, some – but not all – researchers find that pollution and GDP both contain unit roots. If so, shocks have permanent effects on the PIR and previous results are spurious unless pollution and income are cointegrated. Whether or not pollution and income are cointegrated is still open to debate although there is some weak evidence that they are. In this case the country specific results are valid. There is also some evidence that economic growth is bad for the environment unless these effects are corrected by other means.

The main finding is therefore that the EKC is a real world phenomenon, but only for a few pollutants and not in all countries alike. Two natural questions arise: Why is pollution rising at low income levels, but falling at higher ones? And why is there an EKC for some pollutants, but not for others? The reduced form models analysed in this chapter are not able to give an answer. We will tackle these questions in the rest of this book.

3. Possible Causes of the Environmental Kuznets Curve: The Theoretical Literature

3.1 Introduction

In chapter 2 we have surveyed the empirical literature about the PIR. We have found that for some pollutants in some countries an EKC is observed. In this chapter we therefore review the literature to find possible causes of this inverted U-shaped PIR. We also discuss several theoretical models and arguments which predict an EKC. Furthermore, we consider the empirical relevance of these causes. All literature surveys to date concentrate on the empirical literature as we did in chapter 2. Therefore such a detailed survey of all possible causes of the EKC and in particular of their empirical relevance has never been given before. One of the main questions we have to answer in this chapter is whether the EKC occurs automatically in the course of economic growth or whether environmental policy plays a crucial role in causing the downturn of the EKC. The survey might also give some clues why there is an EKC for some pollutants, but not for others.

Theoretical models make many simplifying assumptions about the ecosystem. One of the main assumptions is that different types of pollution can be considered separately. In reality, however, there are many interconnections (*Beltratti* 1996, 2 and *Moslener* and *Requate* 2001). Nevertheless, this assumption is often necessary to make the models tractable. Moreover, in many models dealing with economic growth and the environment the EKC is excluded by assumption: It is assumed that the economy follows a balanced growth path on which pollution stays constant.[1] We will not discuss these models because they are not related to the EKC and because as seen in chapter 2 the evidence rejects the assumption of balanced growth.

All other things equal economic growth is bad for the environment. This is due to the first law of thermodynamics which states that no material can be destroyed. Hence, when economic activities increase, more material is 'transformed into other goods (bound to become waste some time), into waste, and pollution' (*Neumayer* 1998, 162). This so-called *scale effect* explains that pollution increases at low income levels.

In the rest of this chapter we will discuss several counteracting forces which might cause pollution to decline at higher income levels. In section 3.2 we argue that increasing demand for environmental quality can cause the downturn of the EKC. In subsection 3.2.1 we discuss several models which all predict an EKC because of the internalization of external effects by the government. In section 3.3 we show that the EKC for a certain pollutant might emerge because this pollutant is substituted by another pollutant. In section 3.4 we describe that technological progress can cause the EKC. In section 3.5 we introduce the idea that increasing returns to scale in abatement could play a key role. In section 3.6 we consider the view that structural change – the rise and fall of the industrial sector – explains the EKC. In section 3.7 we present the consequences for the EKC when it is caused by the migration of dirty industries from developed into developing countries. In section 3.8 we analyse whether or not the income distribution influences the EKC. In section 3.9 we derive that shocks – mostly caused by the government – could also be important for the EKC. In section 3.10 we state that a falling EKC does not imply that environmental quality rises with income when irreversibilities are

[1] To name a few this assumption can be found in *Bovenberg* and *Smulders* (1995 and 1996), *Smulders* (1995a and 1995b), *van Ewijk* and *van Wijnbergen* (1995), and *Smulders* and *Gradus* (1996).

involved. We therefore argue in subsection 3.10.1 that the fear of overstepping a threshold might cause the downturn of the EKC. Since it is important whether the turning point lies above or below a threshold level, we discuss in subsection 3.10.2 some theoretical results concerning the location of the turning point of the EKC. In section 3.11 we summarize our findings and draw some important conclusions.

3.2 Demand for environmental quality

A first possible cause of the EKC is that demand for environmental quality increases with income, i.e. that environmental quality is a normal good. There are several reasons for this argument: First, only when the basic needs are met, additional resources are devoted to combat pollution. Rising income makes these resources available. Second, with rising income immaterial goods such as environmental quality become more important (*Helm* 1995, 32, *Eglin* 1995, and *Vogel* 1999, 32). Third, the average education increases with rising income. Thus environmental awareness (*Selden* and *Song* 1994), the fear of environmental health hazards, and the concern for the reduced life expectancy grow as well (*Gruhl* 1992, 119). Finally, the opportunity costs of lost work days due to health problems increase with income because wages rise (*Shafik* 1994). In chapter 5 we will show that in a simple model and assuming standard functional forms it is even necessary for the downturn of the EKC that environmental quality is a normal good which means – as we will show – that there is a tendency to satiation in consumption.

It is often argued that environmental quality is a luxury good, i.e. that the income elasticity of demand for environmental quality is greater than one (*Ansuategi et al.* 1998, *Neumayer* 1998, *Vogel* 1999, 76, *Cole* 2000b, 27, and *de Bruyn* 2000, 89). However, the empirical evidence shows that the income elasticity is positive, but smaller than one, i.e. that environmental quality is a normal, but not a luxury good (*Kriström* and *Riera* 1996). Furthermore, it will be shown in chapter 5 that even theoretically there is no unambiguous relationship between the PIR and the income elasticity of demand for environmental quality. Hence, the luxury good argument is not tenable – neither from an empirical, nor from a theoretical point of view.

Since abatement is a public good, the government must ensure that the increased demand for environmental quality can be met. With rising income people increase their support for environmental policies in elections and referenda. They might also join an environmental lobbying group to put the government under pressure – an effective way of bringing down pollution (*Hettige et al.* 1996, 1898). Relatively advanced political institutions are necessary to internalize externalities and to enforce regulations and thus to cause the downturn of the EKC. However, advanced social, legal, and fiscal institutions may only be implementable in rich democratic countries (*Baldwin* 1995, 61, *de Bruyn* 1997, 487, and *Andreoni* and *Levinson* 2001, 270). There is evidence that higher GDP makes policy more environmentally friendly (*Congleton* 1992, *Ringquist* 1993, and *de Bruyn* 1997), that democratic countries have smaller pollution levels (*Torras* and *Boyce* 1998, *Harbaugh et al.* 2000, and *Carlsson* and *Lundström* 2001), and that democracies are more likely to sign international environmental agreements (*Congleton* 1992 and *Fredriksson* and *Gaston* 2000). Further evidence shows that political and civil freedoms significantly decrease pollution of only those pollutants which have direct health effects (*Barrett* and *Graddy* 2000, 455): For these pollutants pressure tends to be highest.

Hence, increasing demand for environmental quality and the ensuing policy measures might actually cause the EKC – at least for pollutants involving high health hazards.

3.2.1 Internalization of external effects

Internalization of external effects as discussed above can also cause an N-shaped PIR: Based on a simplistic model *Pezzey* (1989, 25–29) predicts that, first, pollution is externalized such that pollution grows due to the scale effect (as discussed in the introduction). As pollution and income grow, however, the policy maker starts to internalize external effects. As internalization proceeds, pollution declines. When internalization is complete, pollution starts to increase again because of the scale effect.

In the rest of this subsection we discuss several interrelated models of the EKC. In all these models the EKC emerges because there is no abatement at low income levels, but at higher income levels abatement expenditures become positive because demand for environmental quality increases with income. Since abatement is a public good, policy is again of prime importance. Often these models predict an inverted V-shaped PIR, rather than an inverted U-shaped PIR because pollution suddenly starts to decline when abatement expenditures become positive. All these models assume that a utility function U is maximized which depends on consumption C and pollution P, $U(C, P)$, where $U_C > 0$ and $U_P < 0$.

McConnell (1997) develops a static model in which pollution is generated by consumption whereas in *Selden* and *Song's* (1995) dynamic model pollution is generated by production capital. In both models all external effects are internalized by the government. These two models will be considered in detail in chapters 5–7. In section 5.4 we show that the model of *Stokey* (1998) is a special case of the model of *McConnell* (1997).[2] While *Selden* and *Song* (1995) and *McConnell* (1997) only consider flow pollutants, *Stokey* (1998) extends her model to stock pollutants and also analyses two dynamic extensions of her static model. She shows that the main

[2]This is not completely true as the cross derivative of the pollution function is negative in *Stokey's* (1998) model, but zero in *McConnell's* (1997) model. However, both models are special cases of the general model we will present in chapter 5.

result from the static model – namely that there is an EKC if there is asymptotic satiation in consumption, but that pollution is monotonically rising if there is no satiation (see proposition 5.4 below) – holds also in these three extensions.

The main driving force behind the EKC in these models is a tendency to satiation in consumption which causes demand for environmental quality to increase with income. Since it is assumed that all external effects are internalized, it follows that regulations tighten. Note that if pollution is high or if abatement is cheap, abatement expenditures become positive at a lower income level in these models. *Chimeli* and *Braden* (2001) – whose dynamic model with a stock pollutant yields similar results – additionally show that the growth rate of consumption becomes smaller when abatement expenditures become positive because part of output is then used for abatement instead of consumption.

In an overlapping generations model with a stock pollutant *John* and *Pecchenino* (1994) derive similar results. This model will be discussed at length in chapter 8 where we analyse a model based on the model of *John* and *Pecchenino* (1994). In their model there is also no abatement at low income levels because the country is poor and enjoys a high environmental quality, i.e. a low pollution stock. At higher income levels abatement causes the downturn of the EKC. However, if there are external increasing returns to scale in production, a low capital, high pollution trap might ensue: A country trapped in this steady state invests so much in abatement (because pollution is high) that no resources remain for economic growth (because the country is poor).[3]

[3]We should also mention the model of *Jones* and *Manuelli* (1995). However, their overlapping generations model with a flow pollutant gives unrealistic predictions: For myopic governments *Jones* and *Manuelli* (1995) predict an N-shaped pollution path over time, but pollution is falling because consumption declines which is not sensible. Furthermore, they do not show how income behaves along this path. Actually, pollution and income suddenly jump down at the trough of the N-shape over time. Moreover, the tax revenues are wasted which is again unrealistic. When considering the far-sighted optimum *Jones*

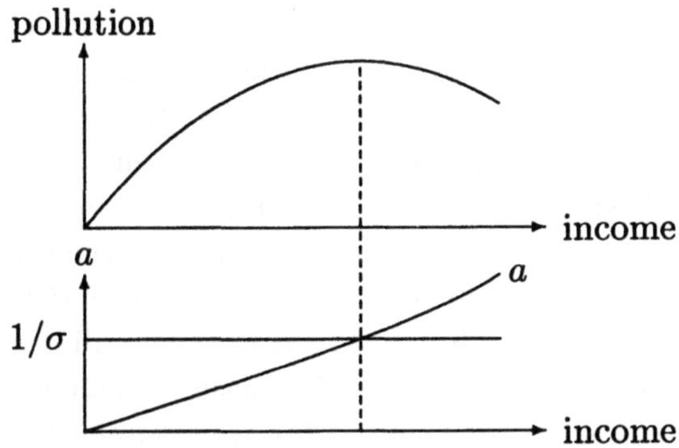

Figure 3.1: The EKC in the model of López *(1994).*

If there is no internalization, pollution rises monotonically with income in the static model of *López* (1994, 165–173). However, if the government adjusts the price of the environment as a factor of production, we find that

$$\text{the PIR is } \begin{matrix}\text{rising}\\ \text{falling}\end{matrix} \quad \text{if} \quad 1/\sigma \begin{matrix}>\\ <\end{matrix} a := -U_{CC}C/U_C \qquad (3.1)$$

where σ is the constant elasticity of substitution in production between conventional inputs (labour and capital) and pollution and where a is the elasticity of marginal utility of consumption which is also called the coefficient of relative risk aversion (*Mas-Colell et al.* 1995, 195). Hence, the higher σ and a, the higher is the probability that pollution declines with income. The intuition behind this result is as follows: Income growth increases the value of the environment to the consumers. Thus the government rises the price of pollution as an input in production. If the marginal utility of consumption falls fast (a high), a large price increase is necessary because demand for environmental quality rises

and *Manuelli* (1995) find an inverted U-shaped pollution path over time caused by zero and positive abatement expenditures. However, *Jones* and *Manuelli* (1995, 6) state that myopia is more realistic.

fast. Furthermore, if substitution between conventional inputs in production and pollution is high (σ high), a given price increase induces a large reduction of the polluting input because there are many possibilities to substitute away from the polluting input. Finally, *López* (1994) shows that under plausible conditions a increases with income while σ is assumed to be constant as shown in figure 3.1. Since $a = 0$ for zero income, the PIR first rises, but when a crosses the threshold level $1/\sigma$, pollution starts to decline: An inverted U-shaped EKC arises. Hence, the EKC is again caused by the rising demand for environmental quality and the assumption that all externalities are internalized.

The models analysed so far consider only the connection from economic growth to environmental degradation, but they neglect that pollution also has an influence on growth. This is possibly the case because 'the effect of economic growth on environmental quality outweighs, both in magnitude and importance, the reversed effect' (*de Bruyn* 2000, 6). Remember also that we could not detect a simultaneity bias in the regression results. Nevertheless, these reversed effects exist: High pollution induces many health problems and thus reduces the productivity of the labour force (*Gangadharan* and *Valenzuela* 2001, 529). Moreover, environmental degradation reduces harvests, forestry yields, and fish catches (*McConnell* 1997, 392). There is evidence that US states with more stringent environmental regulation grow faster (*Carson et al.* 1997, 433). Hence, *McConnell* (1997, 392–393) and *López* (1994, 173–179) extend their models such that higher pollution reduces output. They find that an EKC – caused by zero and positive abatement expenditures and the internalization of external effects – is possible in this case even if consumers do not suffer from pollution ($U_P = 0$) because lower pollution allows for increased output.

Internalization – caused by the rising demand for environmental quality – is central in all these models. Thus they help explain why there is an EKC for air pollutants such as SO_2 whose external costs are at least partly internalized while the PIR is ever rising for CO_2 and municipal waste whose external effects are not yet internalized (*Stern* 1998, 176). In fact, many empirical studies conclude that environmental policy is of prime importance for the downturn of the EKC.[4] Hence, the models discussed in this subsection might capture an important part of the story behind the EKC. However, *Selden* and *Song* (1995) and *Jaeger* (1999) urge caution in interpreting these models. The similarity between observed EKCs and model predictions does not mean that we are following an optimal path. Actual pollution might well lie above or below its optimal level.

3.3 Substitution between pollutants

The downturn of the EKC for one pollutant might also occur because it is substituted by another pollutant. So a flow or a local pollutant could be substituted by a stock, a transboundary, or a global pollutant. If the government regulates only some pollutants, but leaves other pollutants unregulated, the firms substitute away from the regulated to the unregulated pollutants (*Devlin* and *Grafton* 1994). The internalization of a currently existing externality might also result in emissions of a new, hitherto not emitted chemical substance whose consequences on the environment might yet be unknown (*de Bruyn* 2000, 87). The reason can be technological progress. Alternatively, firms might reduce the input of a certain substance due to internalization. If this substance is a joint product of another firm and if it has formerly been fully used as an input to the first firm, it is now released into the environment (*Baumgärtner* and *Jöst* 2000). Furthermore, energy gained from fossil fuels might also be replaced by nuclear

[4]See for example *Grossman* (1995), *Cole et al.* (1997), *de Bruyn* (1997), *Lim* (1997), *Panayotou* (1997), *Hilton* and *Levinson* (1998), *Unruh* and *Moomaw* (1998), *Selden et al.* (1999), *Barrett* and *Graddy* (2000), *de Bruyn* (2000), *Hettige et al.* (2000), and *Minliang et al.* (2001).

power – an energy source with its own problems. This shifts costs on future generations. There is evidence that increased nuclear power generation has helped reduce emissions of SO_2, SPM, and NO_x (*Scruggs* 1998, 271, *Selden et al.* 1999, 14, and *Viguier* 1999).

Finally, abatement technologies themselves cause some pollution while they abate other pollutants (*Huesemann* 2001). This fact is disregarded by the EKC-literature to date although it is often observed in reality as the following two examples show. First, in the catalytic converter of a car the exhaust components – CO, NO_x, and Hydrocarbons – are all transformed into CO_2 (*Holzbaur et al.* 1996, 66). Furthermore, the catalytic converter increases fuel use and thus CO_2 emissions (*Cole* 2000b, 76) and it consists in part of platinum which starts to accumulate in the soil. Second, end-of-pipe technologies have been the most important determinant in causing the downturn of the EKC for SO_2 (*de Bruyn* 1997, 493 and *Ekins* 1997, 822). To run an end-of-pipe technology, however, energy is needed. Since most energy is gained by burning fossil fuels, this causes higher CO_2 emissions. More generally, every end-of-pipe technology consumes energy and therefore increases CO_2 emissions (*Faber et al.* 1996, 272). This might explain why *Stern* (1998, 183–184) has noticed a shift from SO_2 and NO_x emissions to CO_2 emissions.

Hence, substitution between pollutants may well be one of the main driving forces behind the EKC. In part III of this book we will present a model with a similar explanation of the EKC. We will argue that the EKC for a flow pollutant might be due to ever rising stock pollution and due to the neglect of future damages. We will explicitly consider that the abatement technology for one pollutant might have an external effect on another pollutant.

3.4 Technological progress

Technological progress is also important for the pollution path. Note that technological progress is linked to economic growth because with rapid economic growth 'more modern capital is newly installed or replaces old capital' (*Neumayer* 1998, 164). However, *a priori* it is not clear whether technological progress reduces or enhances emissions. Since 'research and development (R&D) efforts by the private sector are still oriented more towards the development of conventional factors saving techniques than to environmental saving techniques', growth tends to increase pollution (*López* 1992, 154).

However, with an appropriate policy technological progress moves in an environmentally friendly direction. So *Nentwig* (1995, 233) and *Stern* and *Common* (2001, 175) state that in the case of SO_2 policy changes induced large amounts of environmentally friendly technological changes. Policy makers also enforced the use of end-of-pipe technologies which have been 'most important' in the reduction of heavy metals emissions (*de Bruyn* 2000, 227). Furthermore, income increases public environmental R&D expenditures (*Komen et al.* 1997) which in turn reduce pollution (*Zaim* and *Taskin* 2000, 33). Hence, policy is of utmost importance for the direction of technological progress.

When a new cheap abatement technology is invented, this technology is implemented in many different countries at about the same time. Since these countries have different income levels, this might explain why the turning point of the EKC varies widely between countries.

Technological progress might also explain the N-shaped PIR. *De Bruyn* (2000, 4) argues that 'reductions in environmental pressure are a temporary phenomenon, that will cease once technological opportunities for further reductions have been exhausted or have become too expensive'. The final upturn of the PIR could therefore 'be explained by the difficulty of keeping up efficiency improvements (innovations) with the continuing growth of production' (*de Bruyn* 2000, 91). Hence, it may be hard to tighten environmental policy – especially when compliance costs become high. Furthermore, 'if there exists a thermodynamic lower bound on reductions in materials and energy use per unit of GDP, emissions may rise once these boundaries are approached' (*de Bruyn* 2000, 91 and *Huesemann* 2001).

Smulders and *Bretschger* (2000) argue that policy-induced technology shifts explain the EKC. They develop a model in which initially all firms produce using a traditional technology causing no pollution. Then a new technology is invented which allows for lower labour costs in production. As more and more firms adopt this new technology, however, pollution rises. This becomes only apparent after some time due to growing evidence of damages found by natural scientists. So public concern rises which culminates in the imposition of an emission tax. In the model this causes pollution to jump down immediately and to stay constant thereafter. However, at some point in time another technology becomes available which is again environmentally clean. Thus while firms adopt this technology, pollution falls and becomes zero when all firms have adopted it.

Thus in the model of *Smulders* and *Bretschger* (2000) the rising branch of the EKC emerges because a labour-saving, but polluting technology spreads out. The downturn of the EKC occurs because of the imposition of an emission tax and because the tax induces the innovation of a new clean technology. There are two crucial conditions for the downturn of the EKC: First, the clean technology must be invented. Second, the tax must be sufficiently high because otherwise it does not pay to adopt the new technology. If one of these conditions is violated for a certain pollutant, the model does not predict an EKC for this pollutant. Note, however, that the model of *Smulders* and *Bretschger* (2000) predicts an inverted U-shaped pollution path over time. It is not shown that income rises along this path. Nevertheless, it seems

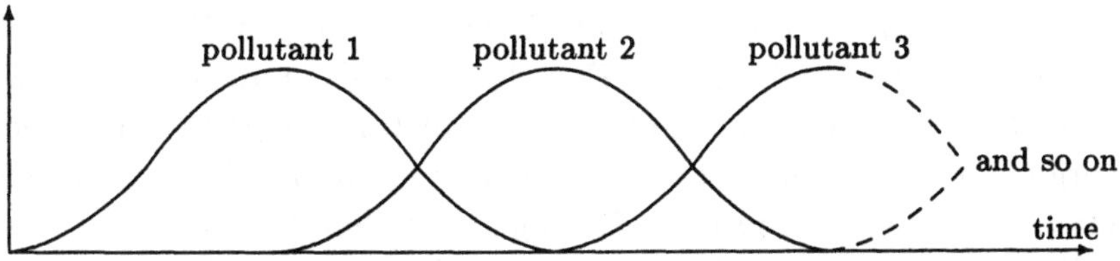

Figure 3.2: Cycles in the model of Smulders *and* Bretschger *(2000).*

plausible that slowly accumulating information about damages and their underlying causes leads to policy measures – but only after damages have become sufficiently obvious: *Wackernagel* and *Rees* (1997b, 9) argue that 'technological fixes limp behind the problems that industrialization causes'.

An extension of the model also allows for cycles as shown in figure 3.2 (*Smulders* and *Bretschger* 2000): After some time it might be realized that a substance emitted by the 'clean' technology also harms the environment. So the first pollutant is only substituted by a second one. Again public concern mounts until a new emission tax is imposed. This hopefully induces another technology such that the second pollutant also approaches zero. Possibly, however, this new technology emits a third pollutant such that a third cycle of the model begins.

Empirical evidence of such cycles for the consumption of thirty important materials in the US is found by *Labys* and *Waddell* (1989). They find that each material follows a life cycle: Consumption first rises and then falls again. However, as consumption of a certain material starts to decline, consumption of other materials begins to rise. Furthermore, due to technological progress we use more and more different materials. Therefore, 'newer materials have been replacing older ones' (*Labys* and *Waddell* 1989, 239). So progress in technological efficiency does not solve all environmental problems as is often thought, but instead typically leads to higher consumption of natural resources since the gains are reinvested in other polluting activities according to *Wackernagel* and *Rees* (1997a, 173 and 1997b, 18–20). This is an indirect way of substitution between pollutants. For instance, despite increased energy efficiency energy consumption increased in OECD countries between 1975 and 1989 (*Wackernagel* and *Rees* 1997b, 18).

Summarizing we find that technological progress is important for the PIR: When abatement technologies or cleaner production technologies become available, growth is possible without increasing pollution. However, only with appropriate market conditions set by policy makers these technologies will actually be invented and used. There is overwhelming empirical evidence backing up the argument that technological progress is crucial for the EKC. This evidence will be discussed in section 3.6 below.

3.5 Increasing returns to scale in abatement

An explanation of the EKC which is also based on technology, but not on technological progress is offered by *Andreoni* and *Levinson* (2001). They argue that increasing returns to scale in abatement can cause the EKC. With increasing returns a larger economy can abate pollution at lower average costs. Hence, a growing economy is able to reduce its pollution in the end. *Andreoni* and *Levinson* (2001) consider a simple static model: Income Y – which rises exogenously – must be split up between consumption C and abatement expenditures A, $Y = C + A$. Pollution P is generated by consumption and can be abated. In particular, they assume $P = \alpha C - R(C, A)$ where αC with $\alpha > 0$[5] represents the emissions and where R are the abated emissions. Finally, utility U is given by $U(C, P)$ where $U_C > 0$ and $U_P < 0$. *Andreoni* and *Levinson* (2001) then prove that in this model an EKC results if the

[5] *Andreoni* and *Levinson* (2001) assume $\alpha = 1$ although their proof goes through for any $\alpha > 0$.

abatement technology $R(C, A)$ exhibits increasing returns to scale (some further conditions are also needed). For a simple example they also show that the PIR is linear and monotonically increasing when the abatement technology exhibits constant returns to scale. This result, however, is only valid if abatement is relatively unproductive. If abatement is relatively productive, in contrast, pollution is always zero.[6]

Andreoni and *Levinson* (2001) also argue that increasing returns are plausible: If an abatement technology involves fixed costs of installment, it has increasing returns because – once these fixed costs are paid for – additional abatement expenditures can be used to actually reduce pollution. Furthermore, for a richer country it may be profitable to use another abatement technology with

[6]The example analysed by *Andreoni* and *Levinson* (2001) is $P = C - C^{\alpha} A^{\beta}$ and $U = C - P = C^{\alpha} A^{\beta}$. The standard Cobb-Douglas solution of $\max_{C,A} C^{\alpha} A^{\beta}$ subject to $Y = C + A$ is $C = Y\alpha/(\alpha+\beta)$ and $A = Y\beta/(\alpha+\beta)$. So when income grows exogenously, we find

$$P = C - C^{\alpha} A^{\beta} = \frac{\alpha}{\alpha+\beta} Y - \left(\frac{\alpha}{\alpha+\beta}\right)^{\alpha} \left(\frac{\beta}{\alpha+\beta}\right)^{\beta} Y^{\alpha+\beta}$$

$$\frac{dP}{dY} = \frac{\alpha}{\alpha+\beta} - (\alpha+\beta) \left(\frac{\alpha}{\alpha+\beta}\right)^{\alpha} \left(\frac{\beta}{\alpha+\beta}\right)^{\beta} Y^{\alpha+\beta-1}$$

$$\frac{d^2P}{dY^2} = -(\alpha+\beta-1)(\alpha+\beta) \left(\frac{\alpha}{\alpha+\beta}\right)^{\alpha} \left(\frac{\beta}{\alpha+\beta}\right)^{\beta} Y^{\alpha+\beta-2}.$$

If the abatement technology has increasing returns to scale ($\alpha+\beta > 1$), the PIR is concave ($d^2P/dY^2 < 0$): An EKC emerges. In contrast, if the abatement technology exhibits constant returns to scale ($\alpha + \beta = 1$), the PIR is linear ($d^2P/dY^2 = 0$). Then if abatement is relatively unproductive, i.e. if $\beta < 0.5$, the PIR is monotonically rising: For $\alpha + \beta = 1$ the slope of the PIR simplifies to $dP/dY = \alpha - \alpha^{\alpha}(1-\alpha)^{1-\alpha}$ which is positive if $1 > \alpha^{\alpha-1}(1-\alpha)^{1-\alpha} = [(1-\alpha)/\alpha]^{1-\alpha}$ or if $1 > (1-\alpha)/\alpha$ which reduces to $\alpha > 0.5$ or $\beta < 0.5$. For $\alpha < 0.5$ the PIR is monotonically falling. However, since the PIR starts at the origin, this means that pollution would be negative. Since negative pollution is not feasible, we are always at the corner solution of zero pollution for $\alpha \leq 0.5$. This has been overlooked by *Andreoni* and *Levinson* (2001, 273) who write that the PIR is always rising for $\alpha + \beta = 1$.

higher fixed costs and lower marginal costs. Looking at some data *Andreoni* and *Levinson* (2001) also find some empirical evidence for increasing returns to scale in abatement. The EKC might also arise because first there is no abatement, but as income grows, 'more and more industries [...] reach the critical size at which the installation costs of abatement equipment can be borne with minimum impact on production costs and profits' (*WTO* 1999, 50).

Hence, in the model of *Andreoni* and *Levinson* (2001) policy does not play any role in causing the downturn of the EKC: The downturn occurs automatically in the course of economic growth due to the increasing returns to scale in abatement. Nevertheless, policy is important: The authors show that without any environmental policy there is also an EKC, but at inefficiently high levels of pollution: So policy should still internalize all external effects. Moreover, in practice, without regulations there might not be any abatement at all.

Finally, since the abatement technologies for different pollutants are different, the model does not predict an EKC for all pollutants. For instance, 'if the abatement technology exhibits increasing returns to scale at low income levels, but decreasing returns to scale at high income levels, an N-shaped path may also result' (*Millimet* and *Stengos* 2000, 3).

3.6 Structural change

Another explanation of the EKC which has been used to justify growth strategies without environmental policies is structural change (*Gangadharan* and *Valenzuela* 2001, 514). People living at the subsistence level produce almost no pollution. However, as the country grows, agriculture intensifies and industry starts to develop causing increasingly more pollution. When the country becomes rich, in contrast, the service sector expands quickly such that pollution might fall again. Hence, the rise and fall of the industry sector could explain the EKC. Furthermore, the structure within the industry sector changes as well: First, 'as income rises, the composition of manufacturing shifts from light to heavy industry', i.e. from relatively clean industries such as food and textiles to polluting industries such as chemicals, minerals, metals, and machinery (*Syrquin* 1988, 242–243 and *Panayotou* 1995). At higher income levels high-tech industries which are far less polluting (*Dinda et al.* 2000, 410) and research activities expand (*Panayotou* 1995).

Structural change can occur because of three reasons (*de Groot* 1999): Because of changing consumer preferences, because of different rates of technological progress in different branches of the economy, and because of policy-induced price increases for polluting inputs in production. *De Groot* (1999) analyses a model in which utility is not reduced by pollution and depends only on the consumption of the goods of different sectors. He assumes that the agricultural sector is characterized by low income elasticities for its goods, by fast technological progress, and by an intermediate emission intensity. In contrast, the service sector exhibits high income elasticities, slow technological progress, and a small emission intensity. The industry sector, finally, is described by intermediate income elasticities, by an intermediate speed of technological progress, and by a high emission intensity. Under these assumptions an EKC emerges due to technological progress and structural change: The agricultural sector shrinks, the service sector expands, and the polluting industry sector shows an inverted U-shaped development. However, *de Groot* (1999) concludes that while structural change helps reduce emissions, it is not sufficient to cause the downturn of the EKC. So emissions-saving technological progress is crucial for the downturn. Without technological progress the EKC only emerges if policy ensures that polluting inputs become increasingly more expensive.

Similarly, in the model of *Cassou* and *Hamilton* (2000) structural change – caused by rising pollution taxes – explains the EKC.

They assume that there are clean and dirty goods and that the dirty goods sector is larger at the outset. Therefore a high tax on the dirty good would reduce consumption considerably because human capital is sector-specific. So the government keeps taxes low and lets the dirty sector grow until the faster growing clean sector becomes sufficiently large. Then the government increases the tax rate faster such that the dirty sector declines again. Thus the dirty sector – and accordingly pollution – follows an inverted U-shaped development path. The main reason for the EKC is that the government wants to smooth the consumption path out. So structural change is policy-induced in the model of *Cassou* and *Hamilton* (2000). Without the optimal policy the PIR would be monotonically rising.

To empirically analyse whether or not structural change is important many researchers have used the estimation equation (2.1) and have added an additional regressor $\beta_5 M_{it}$ where M is the manufacturing share in GDP. While some studies find that β_5 is positive (*Rock* 1996, *Panayotou* 1997, *Borghesi* 2000, and *Cole* 2000a), others obtain that the effect of structural change is only small or even insignificant (*Grossman et al.* 1994, *Grossman* 1995, *Suri* and *Chapman* 1998, and *Auffhammer et al.* 2001b). Moreover, *Cole* (2000a) and *Millimet* (2000) find that structural change alone cannot explain the EKC. Instead, as predicted by *de Groot* (1999) policy and technological progress are crucial for the downturn of the EKC.

Further evidence for structural change and for technological progress comes from decomposition analyses. We first discuss the theoretical underpinnings of decomposition analyses before we turn to the empirical results. The basic idea in every decomposition analysis is that emissions in period t, E_t, can be written as (*de Bruyn* 2000, 164)

$$E_t = \sum_{j=1}^{n} Y_t \, S_{j,t} \, I_{j,t} \tag{3.2}$$

where there are n sectors in the economy, Y is GDP, S_j is the share of sector j in GDP, and I_j is the emission intensity of sector j. Since $S_j = Y_j/Y$ where Y_j is output in sector j and since $I_j = E_j/Y_j$ where E_j are the emissions of sector j, (3.2) is an identity. Differentiating (3.2) with respect to time and dividing by E_t yields

$$\hat{E} = \hat{Y} + \sum_{j=1}^{n} e_j \hat{S}_j + \sum_{j=1}^{n} e_j \hat{I}_j \tag{3.3}$$

where a hat means the percentage change of the variable ($\hat{X} = (dX/dt)/X$) and where $e_j = E_j/E$ is the share of sector j in emissions. The first term on the right hand side of (3.3) is called the *scale effect*, the second term is the *composition effect*, and the third term is the *technique effect*. As discussed in the introduction income growth leads to higher emissions *ceteris paribus* due to the scale effect. The composition effect shows how structural change influences emissions. The technique effect, finally, describes how changes in emission intensities caused by new technologies affect emissions. Some problems arise because (3.3) is in continuous time while actual measurements are only available at discrete points in time. So (3.3) must be transformed into a discrete equivalent (*de Bruyn* 2000, 165). We do not discuss the many different discrete approximations which have been put forward (see *Ang* and *Lee* 1994, *Ang* 1994, and *de Bruyn* 2000, chapter 9).

Decomposition analyses are applied to emissions of SO_2, CO_2, SPM, NO_x, CO, VOC, and heavy metals, as well as to energy and fuel consumption.[7] All studies find that the composition effect is only small or even increases emissions whereas the technique effect is large and always decreases emissions. Hence, decomposition analyses show that the technique effect is much more important for the downturn of the EKC than the composition effect. So we

[7]See *Howarth et al.* (1991), *Torvanger* (1991), *Ang* and *Lee* (1994), *de Bruyn* (1997), *Sun* (1998), *Selden et al.* (1999), *Stern* (1999), *Viguier* (1999), and *de Bruyn* (2000, chapter 11).

again find that structural change cannot explain the EKC (*Selden et al.* 1999). Instead, technological progress is crucial which is likely to be policy-induced as *Smulders* and *Bretschger* (2000, 2) argue.

However, these studies are mostly for rich countries. In poor countries structural change might be an important driving force behind rising pollution. Furthermore, for the reduction of pollutants which cannot easily be abated structural change might play a key role (*de Bruyn* 2000, 218).

Nevertheless, it is often neglected that structural change can only be responsible for the downturn of the EKC because of one reason: The polluting sector must shrink *absolutely* (not only relative to the growing GDP). This is only the case if this sector produces inferior goods whose consumption falls with income – which is 'unlikely' – or if the products of this sector are no longer produced, but imported from other countries (*Torras* and *Boyce* 1998, 149). This will be discussed in the next section.

3.7 Migration of dirty industries

Another possible cause of the EKC is that dirty industries migrate out of rich countries into middle income countries (where the infrastructure is sufficiently developed in contrast to poor countries). Migration thus leads to low pollution in high-income countries and to high pollution in middle-income counties. So migration could explain EKC findings obtained in cross country studies. Since outmigration of dirty industries reduces pollution in rich countries, it can also cause the downturn of the EKC in time series analyses in rich countries. This is alarming for developing countries: While middle-income countries might be able to relocate the dirty industries in even poorer countries, these countries can no longer do so and will face the much more difficult task of abating emissions instead of moving them into other countries.

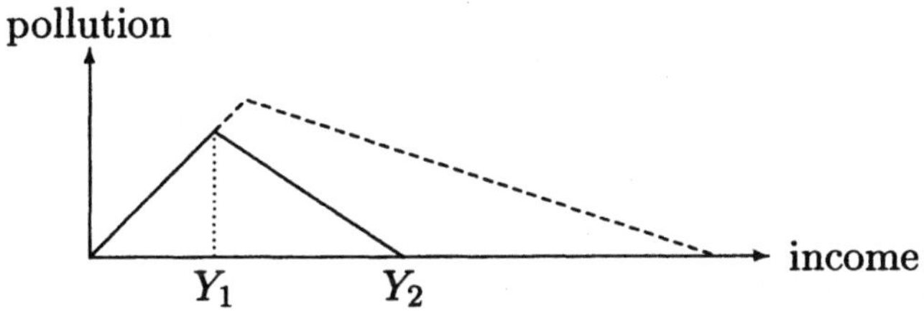

Figure 3.3: The shift of the cross country EKC when the price of the dirty good increases due to world income growth in the model of Saint-Paul *(1995).*

The consequences of migration for the EKC are analysed by *Saint-Paul* (1995). He assumes that there is a dirty and a clean good, that pollution is strictly local and tied to the production of the dirty good, and that the only reason for trade is pollution. He finds that poor countries (income smaller than Y_1 in figure 3.3) produce only the dirty good because its price is higher than the price of the clean good. The richer of two poor countries can produce more and thus suffers from more pollution. A middle-income country produces both goods. The richer of two middle-income countries produces less of the dirty good because the demand for environmental quality rises with income. Finally, a high-income country (income greater than Y_2 in figure 3.3) fully specializes in the production of the clean good and imports its consumption of the dirty good. A cross country analysis thus finds an EKC with zero pollution at high income levels. When all countries grow, they would prefer more consumption of the dirty good, but less pollution. Therefore the price of the dirty good (relative to the clean) rises. Due to this price increase the falling branch of the EKC shifts upwards and to the right (see dashed EKC in figure 3.3) and the rising branch of the EKC becomes longer such that the turning point moves to higher income and pollution levels. Extending the model of *Saint-Paul* (1995) it is easy to show that it is possible that the poorest country is also on the falling branch of the EKC. Furthermore, *Saint-Paul* (1995) shows that the PIR is ever rising in a closed economy. Since the world as a whole is closed, income growth must lead to higher world pollution. So at least part of the countries experience rising pollution – although they all are on the falling branch of the EKC! The reason is that the falling branch of the EKC moves upwards when world income growth induces a price increase of the dirty good.

Looking at the empirical evidence we see that some authors argue that migration of dirty industries is important (*Ayres* 1995, 97), while others think that this effect is too small to be measurable (*Grossman* and *Krueger* 1996, 121). So abatement costs are at most 2,89% (*Tobey* 1990) or 5% (*Ahmed* 1995) of total costs of a firm. However, abatement costs are often underestimated because workplace health and safety protection costs (*Dean* 1992, 19) and energy costs (*Esty* 1994, 22) are not included. Neglecting these costs, *Tobey* (1990), *Jaffe et al.* (1995), and *Jänicke et al.* (1997) find little evidence of migration driven by different environmental policies in rich and poor countries. The literature concentrates on whether or not stricter environmental regulations cause firms to outmigrate (to assess the effect of tighter environmental policies). However, when analysing the EKC it is important that migration could also occur because of other reasons such as differences in labour costs. Nevertheless, migration could lead to falling pollution in rich countries and thus to the observation of an EKC.

Migration can occur in two ways. On the one hand, firms may actually move their production site from developed into developing countries. Examples are industries with high toxic intensities (*Birdsall* and *Wheeler* 1992) and the leather tanning industry (*Ayres* 1997, 423). On the other hand, production sites in developing countries may grow faster than sites in developed ones. In fact, *Low* (1992) shows that dirty industries grow faster in developing than in developed countries, and *Perrings* and *Ansuategi* (2000, 23) find that 'developing countries account for a steadily increasing proportion of world output in many of the most highly polluting industries'. Similarly, *Stern* (1998, 185) observes that the poorest US states often host more polluting industries while the richest states have large service sectors. Hence, the findings of *Carson et al.* (1997) – an EKC for seven different pollutants across US states (see tables 2.1, 2.2, and 2.5) – can possibly be explained by migration. Moreover, *Cole* (2000a) obtains that the ratio of dirty to clean manufacturing output rises at low income levels, but falls at higher ones with a turning point at 5300 PPP$. Hence, there is evidence for migration although migration is possibly not driven by environmental policy.

Further evidence for migration is obtained when one finds that the consumption of dirty goods in developed countries rises faster than their production (*Rothman* 1998). This is possible due to imports from developing countries. The export share of dirty industries is growing in developing countries while it is falling in the world as a whole (*Low* and *Yeats* 1992). Furthermore, *Suri* and *Chapman* (1998) conclude from their empirical analysis that imports help causing the downturn of the EKC. Most EKC studies analyse production based measures of pollution. *Rothman* (1998) argues that consumption based measures should be used instead. So he analyses the ecological footprint – a consumption based measure – and finds a monotonically rising PIR (cf. table 2.5). Hence, the interpretation that the EKC found for production based measures is a reason for optimism might be wrong: Economic growth might not solve environmental problems but might only relocate them. Furthermore, trade itself is polluting: International trade is responsible for one eighth of world oil consumption (*Cole* 2000b, 28). 'Thus trade contributes substantially to energy-related environmental damage' (*Ekins et al.* 1994, 7–8).

Migration is also proposed as an explanation of the finding of an EKC for hazardous waste sites in US counties (see table 2.5). *Gawande et al.* (2000 and 2001) argue that high-skilled high-income workers move out of counties with many hazardous waste sites into clean counties. So according to this theory households are moving, not firms. However, this theory is only useful for strictly local pollutants and if workers are very mobile.

Finally, it has been argued that migration is beneficial because countries with higher assimilative capacities should host more polluting firms (*Dean* 1992, 16, *WTO* 1999, 50, and *Ansuategi* 2000, 47). However, this presupposes that all external effects are internalized in all countries. Since this is not the case, it is important that 'tropical resources such as forests, fisheries and soils [...] are known to be more fragile and less resilient than temperate resources' (*Neumayer* 1998, 170). Hence, the relocation of pollution into developing countries might have serious effects on their environment.

Summarizing we conclude that both the evidence from the development of dirty industries and the evidence from trade and from consumption based measures of pollution show that migration of dirty industries could explain the EKC. This has important consequences. It implies that countries do not move along the EKC as estimated in cross country studies. Instead, the whole cross country EKC is likely to move upwards when income grows. Furthermore, it means that developing countries cannot follow the same path that developed ones have taken (as obtained from time series analyses). Developing countries will actually have to reduce pollution by abating it or by curtailing consumption while developed countries could relocate the polluting production process without having to lower consumption.

3.8 Income distribution

It is argued that the income distribution might influence the PIR. Suppose that environmental policy is determined by the median voter[8] and that environmental quality is a normal good (as we argued in section 3.2). Then a more equal income distribution

[8]'The median voter theorem is widely applied [...] to so-called general interest issues' such as the environment (*Eriksson* and *Persson* 2001, 6).

makes the median voter better off and therefore environmental regulation tightens and pollution falls. So the EKC might be caused by the original Kuznets curve which states that income inequality – instead of pollution – follows an inverted U-shaped path when income grows. However, for this to be true, the income of the median voter must actually decline at low income levels although average income rises. Due to the deepening inequality this is possible, but it might seem implausible. Nevertheless, falling inequality at higher income levels might help reduce pollution.

However, reviewing the literature *Vogel* (1999, 99–100) finds that the richest households are not most in favour of strict environmental regulations. In fact, the middle class – earning 50 to 200% of the median income – is most supportive to strict environmental policies (*Vogel* 1999, 100). There are two possible explanations for the green middle class. First, it might be due to different sources of income (*Vogel* 1999, 101–106): The unskilled, badly-paid workers of extractive or heavy industries must often fear to loose their jobs when environmental regulations are tightened. In contrast, middle class employees work mostly in the clean service sector and are thus free from such fears. The upper class, finally, derives a substantial part of income from profits which tend to fall with stricter environmental policies. Therefore, despite their higher income, the rich prefer a laxer policy in order to increase their profits and incomes. The second explanation of the green middle class is that wealthy households might be able to improve their environmental quality by private measures (*Vogel* 1999, 106–112 and *McConnell* 1997, 393): The rich can afford to live in a clean neighbourhood, to install air conditioning, and to drink bottled water instead of tab water. Once these measures have been taken, the rich are no longer willing to pay for improvements in public environmental quality.

Since the middle class is green, we again find that the original Kuznets curve might cause the EKC: The middle class is smallest at intermediate income levels where income distribution is most unequal and pollution highest. However, *Vogel* (1999, 113) argues that it seems more likely that the development of the industry sector causes both, high income inequality and high pollution. But again, a more equal income distribution might help cause the downturn of the EKC. Or put differently, 'large distributive imbalances in high-income countries can prevent environmentally friendly policies from being adopted and pollution from declining' (*Magnani* 2000b, 14).

A similar conclusion is drawn by *Torras* and *Boyce* (1998). They state that if income distribution is highly unequal, the poor who suffer from pollution do not have much political power. Instead, the wealthy who derive profits from polluting activities dominate the political process. Hence, a more equal income distribution again leads to stricter environmental policies. In contrast, *Gangadharan* and *Valenzuela* (2001, 524) argue that due to more equal standards of living 'more people are able to afford the use of electricity, cars, and other luxuries – which leads to increased energy use' and thus to higher pollution.

The empirical evidence for income inequality is rather weak. The Gini coefficient often has the expected sign, but is frequently insignificant (*Scruggs* 1998, *Torras* and *Boyce* 1998, *Borghesi* 2000, *Magnani* 2000a, and *Gangadharan* and *Valenzuela* 2001). This might possibly be due to the poor quality of the data for the Gini coefficient (*Torras* and *Boyce* 1998). Results obtained from using political rights and civil liberties are more supportive to the hypothesis that higher inequality in political power – not in income – aggravates pollution (*Torras* and *Boyce* 1998, *Barrett* and *Graddy* 2000, *Harbaugh et al.* 2000, and *Carlsson* and *Lundström* 2001).

Summarizing we conclude that empirically the income distribution has only a small effect on the PIR. Inequality in political power might be more important than income inequality. An equal distribution of political power possibly helps causing the downturn of the EKC.

3.9 Shocks

In section 2.4 we have shown that shocks have permanent effects on the PIR if the income or pollution data contain unit roots and are not cointegrated. As we have seen the empirical evidence is inconclusive. This leaves the possibility open that shocks are at the root of the EKC. Then the future path of pollution cannot be predicted (*Labson* and *Crompton* 1993). So *Moomaw* and *Unruh* (1997) argue that the oil price shock might have been more important for the turning point of the EKC for CO_2 than income: In 16 OECD countries the turning point has occurred in the 1970s at widely different income levels (8900–15400 PPP$). However, income is still important: The oil price shock has only led to declining CO_2 emissions in the wealthy OECD countries. *Unruh* and *Moomaw* (1998, 227) state that shocks overcome the political and economic inertia and cause changes which often occur within a single year. In contrast, *de Bruyn* (2000, 124) obtains that changes in energy prices do not have a significant effect on changes in pollution. Moreover, the time dummies in the estimation of *Stern* and *Common* (2001, 171) do not point to sudden declines due to the oil price shock (or other shocks). Hence, if the oil price shock has an effect, it becomes only gradually visible.

De Bruyn (2000, 152) argues that when an economy is in equilibrium the PIR is monotonically rising due to the scale effect. However, shocks can lead to reductions in pollution. Once the effects of the shock evaporate, the economy reaches a new equilibrium and thus pollution rises again with income. Hence, an N-shaped PIR results. There is evidence of such behaviour for steel and energy consumption in four OECD countries (*de Bruyn* 2000, 152–156). Hence, institutional and technological breakthroughs are required to keep pollution at reasonably low levels when income grows (*de Bruyn* 2000, 156).

The EKC might also be a spurious result because there is a structural break: Developing countries might follow a monotonically rising PIR while the PIR of developed countries might be monotonically falling (*Vincent* 1997). *Halkos* and *Tsionas* (2001) try to estimate such behaviour. However, for CO_2 and deforestation they find that developing and developed countries follow almost the same PIR: There is no structural break. Furthermore, evidence appeared for an EKC in a single developing country (China) for waste gas emissions (*Auffhammer et al.* 2001a and 2001b) and for waste water emissions (*Minliang et al.* 2001).

Since shocks possibly explain the EKC (see inconclusive evidence in section 2.4 and here), it is important to know the nature of these shocks. Shocks might result from price changes (oil price), from policy measures (*Unruh* and *Moomaw* 1998, 227), or from technological innovations (*de Bruyn* 2000, 151). The oil price shock was the result of a policy measure of the OPEC (organisation of the petroleum exporting countries). Price changes also result from the internalization of external effects. Furthermore, technological innovations might also be influenced by policy measures. Hence, with the possible exception of technological innovations all shocks go back to policy measures. Thus rising demand for environmental quality and rising scientific knowledge about the environmental consequences of production might lead to sudden changes in policy. Therefore, shocks are not a new explanation of the EKC. However, internalization is no longer considered as a continuous process, but as a succession of sudden jumps.

3.10 Irreversibilities

Before concluding this chapter we must discuss a crucial assumption behind the EKC whose violation might lead to another cause of the EKC. It is always assumed that falling emissions or concentrations imply that environmental quality improves. The EKC seems to signify that environmental quality is equal on the rising and on the falling branch of the EKC at the same level of emissions or concentrations. However, this need not be the case. As

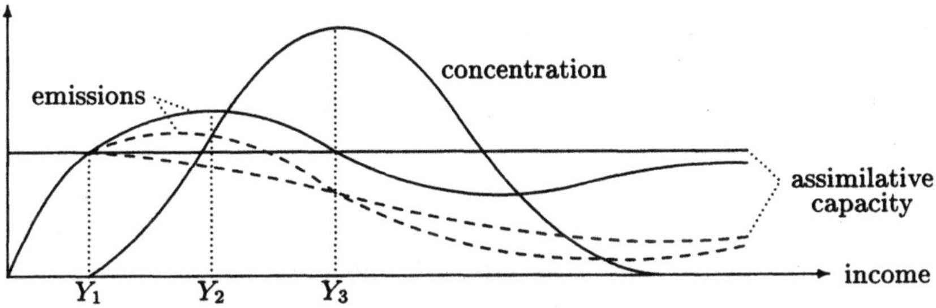

Figure 3.4: Emissions, the concentration, and the assimilative capacity.

mentioned in subsection 2.3.2 even if emissions per capita fall, total emissions might rise because the population grows. Furthermore, the most obvious (but extreme) counterexample are irreversible damages. When a certain threshold level of emissions or concentrations is overstepped, an irreversible damage might occur. Then falling emissions or concentrations do not mean that environmental quality improves again. In fact, even zero emissions or concentrations are not sufficient to restore the pristine state of the environment (*Vogel* 1999, 22). So even if an EKC for soil erosion, desertification, extinction of animal and plant species, biodiversity loss, and nuclear waste generation exists, the involved damages are irreversible. For instance, when a piece of land is cleared of trees and then reafforested, we again have the same area of forest, but the biodiversity is much lower (*Nentwig* 1995, 550).

A similar effect can also be found in a less extreme example. Suppose that nature can assimilate a certain amount of emissions. However, when emissions of a steadily growing economy exceed the assimilative capacity, the pollutant starts to accumulate (at income level Y_1 in figure 3.4). This concentration might for example affect human health. Only when emissions fall again below the assimilative capacity, concentration begins to decline and hopefully approaches zero as in figure 3.4 (*Sprösser* 1988, 77–83). For income levels between Y_2 and Y_3 emissions decline, but the concen-

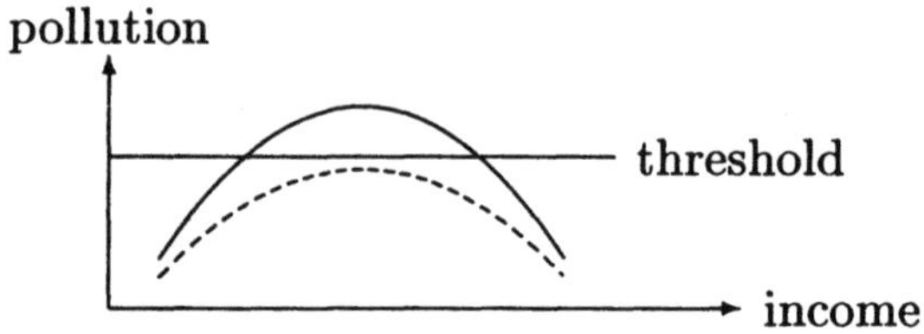

Figure 3.5: Tunneling through the EKC.

tration still increases and causes increasingly higher human health problems. It might be that the concentration also reduces the assimilative capacity (see dashed line in figure 3.4). Then the concentration follows only the same path if emissions are immediately reduced accordingly (dashed path). Again if emissions fall below the (declining) assimilative capacity, the concentration starts to decrease, but the assimilative capacity still declines, i.e. the state of the nature still worsens. Hence, we also find that falling concentrations along the EKC path are no guarantee that the state of the environment improves. In this example the damage might be reversible: When concentrations are back to zero, the assimilative capacity might recover and rise back to its original level. However, the costs of reversing the damage are high since emissions must be reduced drastically as they must fall below the declining assimilative capacity.

3.10.1 Tunneling through the environmental Kuznets curve

Therefore it is argued that policy makers in countries on the rising branch of the EKC should ensure that thresholds are not overstepped and irreversibilities are prevented – or as *Munasinghe* (1995 and 1999) and *Panayotou* (1997) have called it, governments should 'tunnel through the EKC' (see figure 3.5). Through appropriate policy measures – such as the removal of energy subsidies, the introduction of more secure property rights over natural resources, and the internalization of externalities (*Panayotou* 1997, 469) – the EKC becomes flatter such that the turning point of the EKC lies at a smaller pollution level. Hence, as mentioned in subsection 2.3.2, the pollution level at the turning point of the EKC is more important than the income level because the pollution level determines whether or not (irreversible) damages occur (*Agras* and *Chapman* 1999, *Borghesi* 1999, and *Perrings* and *Ansuategi* 2000). Since we have found that each country follows its own EKC (see paragraph 2.3.3.6), developing countries could follow a flatter EKC than developed countries (although the migration of dirty industries might make this a difficult task).

Tunneling through the EKC might imply that economic growth becomes slower (*Munasinghe* 1999). Nevertheless, this might be the wiser development strategy because otherwise money must be spent at higher income levels 'on extremely costly, time-consuming, and often ineffectual assessment, cleanup, and restoration activities' (*Schindler* 1996, 18). The problem, however, is that thresholds are not known (*Neumayer* 1998, 170). Furthermore, the interrelation between different pollutants are frequently unknown such that their joint effect on the environment remains in the dark. This often leads to an 'underestimation of their true harmfulness' (*Vogel* 1999, 182).

Nevertheless, the fear of transgressing a threshold – which is not precisely known – can increase the demand for environmental quality and thus cause the EKC. It might seem that this fear can only explain why pollution stops increasing, but not why it turns down. However, to prevent concentrations from increasing, emissions must possibly fall. This implies an EKC for emissions (cf. figure 3.4). If the concentration causes damages, it might also be necessary to decreases the concentration in order to prevent further damages. For instance, considering greenhouse gases the fear of large anthropogenic climate changes has led to the Kyoto protocol. Thus the world is about at an income level of Y_2 in figure

3.4 where emissions start declining but where the concentration continues to rise.

3.10.2 The location of the turning point

Three results discussing the location of the turning point of the EKC might be important to determine whether the turning point lies below or above a threshold level, i.e. whether or not we succeed in tunneling through the EKC. *López* and *Mitra* (2000) develop a model to show that in a more corrupt country the turning point occurs at a higher pollution and income level. With corruption pollution is always higher than in the social optimum because polluting firms give lobby-payments to the government who pays back with lax environmental policies. The more corrupt the government is, the higher is pollution. Thus to tunnel through the EKC it might be crucial to combat corruption. However, it is unlikely that growth reduces corruption rapidly because 'institutions and cultural norms typically show extraordinary resilience' (*López* and *Mitra* 2000, 149). Hence, fast growing developing countries with high corruption levels – such as China, India, and Indonesia – tend to reach higher pollution levels than suggested by EKC estimates obtained from relatively corruption-free developed countries (*López* and *Mitra* 2000, 149–150).

A second result deals with transboundary pollutants. Based on an extension of *Selden* and *Song's* (1995) model and assuming non-cooperative behaviour of several identical countries *Ansuategi* and *Perrings* (2000) find that both the level of income and pollution at the turning point of the EKC are higher when the pollutant is more transboundary, i.e. when more emissions are transported to other countries. The reason is that the incentives to abate a transboundary pollutant are smaller because part of the benefits of abatement accrue to foreign countries. If a high percentage of emissions can be externalized to other countries, the PIR is actually monotonically rising. *Ansuategi* (2000, chapter 5) finds some evidence to back up this theory, but he does not analyse whether or not the relatively small (4700–6600$, cf. table 2.1) differences in turning point estimates are significant.

A third result considers intergenerational externalities. Due to selfish, myopic behaviour the turning point of the EKC also lies at higher pollution and income levels when a greater part of damages can be externalized to future generations (*Ansuategi* 2000, chapter 4). Although intuitive, this result – based on an extension of the overlapping generations model by *John* and *Pecchenino* (1994) – rests on many restrictive assumptions.

The last two theoretical results – namely that the turning point lies at higher pollution and income levels when the damage can be externalized to other countries or future generations – underpin and explain empirical findings. They show that international cooperation and far-sighted behaviour are crucial to tunnel through the EKC of transboundary and long-lived pollutants – or to ensure that the PIR of these pollutants is not monotonically rising in the first place. However, the Helsinki Protocol of 1985 and the Sofia Protocol of 1988, which deal with the emissions of two transboundary pollutants – sulphur and NO_x – in Europe, seem not to have been very effective in fostering cooperation: When the protocols were signed, many of the signatory countries had already reached their reduction targets or were soon to reach them (*Murdoch et al.* 1997). Hence, these protocols made only binding the reductions which the countries had already reached on their own because in Europe 'the majority of nations receive over 50% of their own sulphur emissions as depositions' (*Ansuategi* 2000, 20).

In contrast, some progress has been made with the Kyoto Protocol on greenhouse gases. However, most studies find a monotonically rising PIR for CO_2 emissions (see table 2.2). This is probably because the newest data which have been used in empirical studies are those of 1996, but most researchers use older data. Hence, the reductions which should take place until 2010 are not yet visible in the EKC studies to date.

Therefore it could be argued – and we will come back to this argument in the conclusion of this chapter – that *an EKC only emerges for pollutants for which policy measures have been taken, but that for other pollutants the PIR is monotonically rising when the regulator remains inactive.* Hence, regulation is 'a necessary condition for a downward sloping' PIR (*Kelly* 2000, 11). Another good example are chlorofluorocarbons (CFCs): The cross country PIR for CFCs is monotonically rising in 1986 before the signing of the Montreal Protocol in 1987, but it is an EKC in 1990 (*Cole et al.* 1997). After 2006 the cross country PIR for CFCs should be horizontal at zero emissions as CFC emissions are banned after 2006 in developed and developing countries (*Houghton* 1997, 39). This led *Cole et al.* (1997, 412) to conclude that 'without [multilateral action] global air pollutants will increase monotonically with income'. The Montreal Protocol allows developing countries to tunnel through their EKC for CFCs (*Cole* 2000b, 114).

For local pollutants this argument might explain why different countries follow different PIRs (as found in paragraph 2.3.3.6): Each country has its own environmental policy. The argument is also in line with the finding of chapter 2 that local and flow pollutants follow an EKC, but that global and stock pollutants exhibit an ever rising PIR: Flow and local pollutants are more easily internalized than stock and global pollutants (*Cole* 2000b, 24 and 66) because the benefits of abatement are immediately discernible and accrue fully to the local population.

The government only tackles a certain pollutant if this pollutant has received considerable public attention. Otherwise regulations would not be politically feasible (in a democracy). Hence, *without public concern no measures are taken* either by the government or by private actors *to reduce pollution* and there are no incentives to invent cleaner technologies, and pollution will increase with income. Public concern, in turn, can only arise when natural scientists are able to track down the source of a certain damage, i.e. when a substance is revealed to be a pollutant.

In this section we have shown that the falling branch of the EKC is not necessarily equivalent with rising environmental quality, in particular not if there are irreversible damages. Thus the fear of overstepping a threshold is another possible reason for a policy-induced downturn of the EKC. This has led to the view that policy makers should tunnel through the EKC. In fact, it can be argued that an EKC arises only for pollutants which have received considerable public attention and have therefore been tackled by the regulator, but for those pollutants which have not been looked after by the government the PIR is expected to be ever rising.

3.11 Conclusion

In this chapter we have reviewed the many possible causes of the EKC from a theoretical point of view and we have assessed their validity by looking at some further empirical results. Before summarizing the causes of the EKC note that we have also found five different explanations of the N-shaped PIR. The final upturn might occur because once internalization is complete, the PIR is upward sloping, because abatement opportunities are exhausted, because there is a thermodynamic lower bound on material and energy use per unit of GDP, because at higher income levels the abatement technology exhibits no longer increasing, but decreasing returns to scale, or because the downturn is caused by a shock and the rising path is an equilibrium relationship.

Economic growth leads to more output and thus to higher pollution *ceteris paribus*. This scale effect can, however, be counteracted by several forces. There is empirical evidence that demand for environmental quality increases with income. Among other reasons this demand might rise due to the fear of incurring irreversible damages. Provided that relatively advanced political institutions exist, higher demand for environmental quality causes increasingly stricter environmental regulations. This is actually the result of several theoretical models which assume perfect internalization of external effects. In a less perfect world internalization makes progress in the course of economic growth. Regulations may also tighten because pollution reduces output. Hence, the EKC is explained by zero and positive abatement expenditures. The evidence suggests that the income distribution is not crucial for the strictness of policy measures. Inequalities in political power seem to be more important than income inequalities. A more equal distribution of political power might help causing the downturn of the EKC. The internalization of external effects might be a continuous process or it might occur in sudden jumps (shocks).

Internalization can cause technological progress to move in an environmentally friendly direction. So the policy-induced introduction of end-of-pipe technologies has been important for the reductions of several air pollutants. This is corroborated by decomposition analyses which show that changing technologies have been crucial for the downturn of the EKC. However, when appropriate policies are missing, technological progress might well be pollution-enhancing. So *Smulders* and *Bretschger* (2000) propose that pollution increases due to a new technology, that after pollution has raised public concern a tax is imposed, and that this tax induces the invention of a clean technology which allows the downturn of the EKC. However, later it might become clear that a substance emitted by the 'clean' technology is also polluting. This opens up the possibility of cycles.

These cycles mean that one pollutant is substituted by another. So a regulated flow or local pollutant could be substituted by a stock or global pollutant which is not regulated. In fact, end-of-pipe technologies have exactly this effect. Nuclear power has also replaced energy gained from fossil fuels. Substitution might not only occur between two pollutants, but also between two countries: Dirty industries might migrate from developed into developing countries. There is evidence supporting this view. Hence, rich countries can reduce pollution by relocating production and by importing the goods. This causes serious problems as soon as the developing countries want to lower pollution because they cannot relocate pollution but must actually reduce it which is more costly. It has also been shown that the cross country EKC shifts upwards with economic growth when the EKC is based on migration. Therefore a country on the falling branch of the cross country EKC might well see its pollution rise in the course of economic growth due to this upward shift of the EKC.

It has been argued that structural change – the rise and fall of the heavy industry – is 'the most common explanation' of the EKC (*Millimet* 2000,1). However, the evidence is only weak. Structural change can explain the rising branch of the EKC in poor countries, but to cause the downturn the absolute production level of an industry must shrink. This is only plausible when industries migrate. The evidence suggests that structural change alone cannot explain the EKC. Therefore the migration of dirty industries is mostly not so fast that the absolute level of production of these industries declines considerably in developed countries. Nevertheless, dirty industries might grow fast in developing countries. Hence, the migration of dirty industries might explain the cross country EKC, but it cannot explain the EKC of a single, growing developed country over time.

Finally, increasing returns to scale in abatement might allow rich countries to abate at lower average costs than poor countries. This possibly brings about the EKC.

Hence, we have found four empirically validated explanations of the EKC of a growing country over time: Increasing demand for environmental quality, technological progress, substitution between pollutants, and increasing returns to scale in abatement. All these causes could be part of the explanation. A certain cause might be important for one pollutant, but not for another. Several causes might also work together: For instance, growing demand for environmental quality might lead to tighter policies which in turn induce producers to (invent and) use other technologies. These technologies might emit other unregulated pollutants.

Satisfying the increasing demand for environmental quality requires policy measures, technological progress is only environmentally friendly if appropriate incentives are set by policy makers, and substitution is caused by regulations or technological progress. Only the EKC caused by increasing returns to scale in abatement does not directly rely on policy. However, without internalization the EKC lies at inefficiently high levels of pollution. Moreover, without policy measures there might not even be any abatement. So policy is also crucial here. Hence, we conclude that environmental policy is of utmost importance for the downturn of the EKC: *For a certain pollutant an EKC only exists when policy measures are taken but the PIR is ever rising if the government remains inactive.* Therefore fighting corruption might be important since corruption may postpone the implementation of environmental policies. Moreover, since every country has its own environmental policy, this explains why each country has its own PIR.

Nevertheless, 'there appears to be an increasing reliance on the EKC to address environmental degradation' (*Li* 1999, 147). 'Economic growth is still generally seen as the dominant policy objective' although the resilience of the natural system may be threatened (*Common* 1995, 101). However, as we have shown the idea that economic growth is good for the environment is not tenable. *Ayres* (1995, 97) even writes that this idea is 'false and pernicious nonsense'. The key role, in contrast, is played by environmental policy because 'the strongest link between income and pollution in fact is via the induced policy response' (*Grossman* 1995, 43) such that 'the existence of an EKC could be solely the effect of environmental policy' (*de Bruyn* 1997, 499). Hence, 'in the absence of carefully targetted policies [...] economic growth may be just as harmful as its strongest critics suggest' (*Ansuategi et al.* 1998, 158). Many other authors[9] also conclude that policy

[9]See *Barbier* (1997), *Cole et al.* (1997), *Lim* (1997), *Panayotou* (1997), *Hilton* and *Levinson* (1998), *Unruh* and *Moomaw* (1998), *Selden et al.* (1999), *Barrett* and *Graddy* (2000), *Cole* (2000b), *de Bruyn* (2000), *Hettige et al.*

is of prime importance for the downturn of the EKC. To solve transboundary or global pollution problems multilateral or global agreements are therefore needed (*Cole* 2000b, 130).

However, in most cases emission reductions are due to *local* (in space and time) environmental policy reforms (*Arrow et al.* 1995, 92 and *Huesemann* 2001, 283). 'But such reforms often ignore international and intergenerational consequences' because the incentives to take them into account 'are likely to be weak' (*Arrow et al.* 1995, 92). Hence, we find an EKC for flow and local pollutants, but a monotonically rising PIR for stock and global pollutants. Furthermore, policy 'adjustments tend to be made [...] as a belated response to evidence of environmental degradation' (*Perrings* and *Ansuategi* 2000, 34) or even of 'severe environmental degradation' (*Schindler* 1996, 17–18). This is possibly the case because policy makers respond to public concern which arises when scientific knowledge about the sources of these damages becomes available and when the damages are (clearly) visible and because the process of finding acceptable regulations is time-consuming. Nevertheless, it has been argued that policy makers should tunnel through the EKC before severe or even irreversible damages have occurred. This is important because due to irreversibilities even the falling branch of the EKC does not necessarily imply that environmental quality is improving with rising income.

(2000), *Kelly* (2000), *Smulders* and *Bretschger* (2000), *Minliang et al.* (2001), and *Skonhoft* and *Solem* (2001).

Part II

The Environmental Kuznets Curve and Satiation

4. Cardinal Utility in Static and Dynamic Optimization Problems

4.1 Introduction

In part II of this book we scrutinize one of the empirically validated causes of the EKC. We analyse why demand for environmental quality rises with income and find that a tendency to satiation in consumption might be crucial. Before doing so, however, we consider some fundamental points concerning the use of cardinal and ordinal utility functions in static and dynamic models. The methods we develop in this chapter will be employed in chapters 5–8, but they are of a very general nature such that they can also be applied in many other models. In the case of a static model my opinion is that our method is clearly advantageous and should therefore always be used.

An ordinal utility function allows us to say which of two consumption bundles we prefer or whether we are indifferent. The ordinal utility function can be represented by a set of indifference curves: An example is shown in figure 4.1 on page 104. A cardinal utility function also allows us to say by how much we prefer one consumption bundle over the other. It therefore enables us to label all indifference curves with the utility level we derive at each point of this indifference curve. A monotonic transformation of the utility function leaves the indifference curves and thus the ordinal utility function unaffected, but the cardinal utility level of all indifference curves changes.

Hence, assumptions about cardinal utility functions are stronger than assumptions about ordinal utility functions. Therefore, whenever possible we should use ordinal assumptions only. In static models all results can be derived with an ordinal utility function only. Nevertheless, we usually find assumptions about the sign of the second derivatives of the utility function. As we will show in section 4.2 these assumptions are cardinal and thus stronger than necessary.

In section 4.2 we offer a simple method how to derive results from a static model with an ordinal utility function only. The method has four additional advantages: First, it ensures that all goods are normal goods (or, more precisely, it classifies all goods as normal or inferior goods). Second, it has a straightforward graphical interpretation. Third, it allows to simplify equations by writing results in a more compact manner. Fourth, the shorter equations also prevent (or disclose) interpretations which violate the normal good assumption. This is important because with the standard method of using a cardinal utility function such interpretations have been drawn.

We will concentrate on a utility function in which two variables enter. However, our method can also be applied to utility functions depending on more than two variables. Although the method we will present is simple, we have not been able to find a model using our method. We will apply the method in chapters 5 and 8. I strongly believe that the method is of relevance to everyone who wants to develop a static model. Assumptions about the cardinal utility function are not necessary in static models and should therefore not be made.

When considering a dynamic model in section 4.3, however, matters become more complicated. As usual we assume that the utility function is additively separable. Thus a felicity function (also called an instantaneous utility function) exists. In a static model all results are robust to monotonic transformations of the utility function because all results rely only on the ordinal utility function. However, when we transform the felicity function of a dynamic model monotonically, we change the ordinal utility function. In fact, we will show that a monotonic transformation of the felicity function is a change in the elasticity of intertemporal substitution. Therefore at least part of the results change with a monotonic transformation. It is interesting to analyse which results are robust to changes in the elasticity of intertemporal substitution. In subsection 4.3.1 we develop a simple method how we can figure out which results are robust. This method is similar to the method developed for static models. We again concentrate on a felicity function in which two variables enter although our method can also be applied to felicity functions depending on more than two variables. We will employ this method in chapters 6 and 7.

This chapter is organized as follows. In section 4.2 we argue that cardinal utility functions should be banned from static models and we offer a simple framework how this can be done. In section 4.3 we show that monotonic transformations of the felicity function are changes in the elasticity of intertemporal substitution and based on this insight we develop a method how we can check which results are robust to changes in the elasticity of intertemporal substitution. Section 4.4 concludes.

4.2 Banning cardinal utility from static models

In this section we analyse assumptions about the utility function in static models. We suppose that utility U depends on consumption C and pollution P, i.e. $U = U(C, P)$. The points we will make in this chapter can also be derived if utility depends on two goods, instead of on a good and a bad. However, since we will use the example of $U(C, P)$ throughout the rest of this book, we present the basic insights in this chapter also in terms of $U(C, P)$. An example of some indifference curves is shown in figure 4.1. There might (but need not) be a satiation level of consumption which for zero pollution is at C^* (satiation will be of some importance in chapter 5, but we will already use it in chapter 4). We assume $U_C > 0$ (below the possible satiation level) and $U_P < 0$. Moreover, consumption C and environmental quality or $-P$ are normal goods: If the consumer could trade consumption

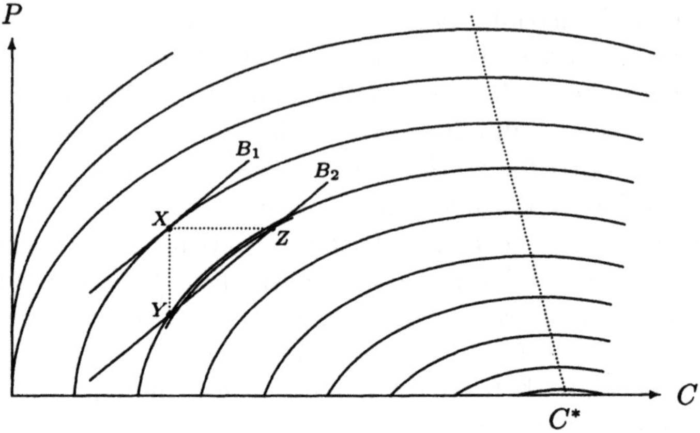

Figure 4.1: Graphical illustration of the normal good assumption.

goods and pollution at a fixed relative 'price' and if his income increased, he would choose more consumption and less pollution. Starting at the optimal point X on the budget line B_1 in figure 4.1 the new optimal point must lie between Y and Z on B_2. In other words, the slope of the indifference curve at Y must be steeper than at X, but flatter at Z than at X. The slope of the indifference curve, i.e. the *marginal rate of substitution* (MRS), is given by $MRS = -U_C/U_P > 0$. Taking the derivative with respect to C and P the normal good assumption means[1,2]

$$MRS_C = \frac{\partial MRS}{\partial C} = \frac{-U_{CC}U_P + U_{PC}U_C}{U_P^2} < 0 \quad (4.1)$$

$$MRS_P = \frac{\partial MRS}{\partial P} = \frac{-U_{CP}U_P + U_{PP}U_C}{U_P^2} < 0. \quad (4.2)$$

[1]Note that these two assumptions also ensure that the utility function is quasiconcave: Moving along an indifference curve to the right C and P increase such that – according to (4.1) and (4.2) – the slope of the indifference curve becomes flatter. Hence, the indifference curves are concave.

[2]If we assumed that C or $-P$ was an inferior good, then MRS_P or MRS_C would be positive.

These are only ordinal assumptions about the utility function since we consider only the shape of the indifference curves, but not cardinal utility levels. All additional assumptions about second derivatives are cardinal. Nevertheless, several authors assume $U_{CC} < 0$, $U_{PP} < 0$, and often also $U_{CP} \leq 0$ or even $U_{CP} = 0$.[3] Whereas $U_{CC} < 0$ and $U_{PP} < 0$ seem to be relatively plausible, $U_{CP} = 0$ is a restriction on allowed preferences (for example, if $U_{CP} = 0$, the satiation level must be at C^* for all pollution levels which excludes preferences as in figure 4.1). Furthermore, with a monotonic transformation we can change a cardinal utility function with $U_{CP} < 0$ (or $U_{CC} < 0$ or $U_{PP} < 0$) into another cardinal utility function – leaving the indifference curves unaffected – with $U_{CP} > 0$ (or $U_{CC} > 0$ or $U_{PP} > 0$) and thus $U_{CC} < 0$ and $U_{PP} < 0$ (or $U_{CP} < 0$) according to (4.1) and (4.2).

Hence, when we derive results from a static model, we should always use MRS_C and MRS_P instead of assumptions about the sign of second derivatives. This is the crucial point of our method. This procedure has two additional advantages: First, we can write equations in a shorter, more compact way using MRS_C and MRS_P. For instance, suppose that income Y is split up between consumption and environmental quality Q, i.e. $Y = C + \pi Q$ where π is the relative price of environmental quality. Then the income elasticity of demand for environmental quality can be written as $Y/Q \; dQ/dY$ where[4]

$$\frac{dQ}{dY} = \frac{MRS_C}{\pi MRS_C + MRS_P}. \tag{4.3}$$

[3]While *McConnell* (1997) realizes that U_{CP} might be positive or negative and both *John* and *Pecchenino* (1994) and *Ansuategi* (2000, chapter 4) assume $U_{CP} \leq 0$, the majority of researchers assumes $U_{CP} = 0$, in particular *Stokey* (1998) whose model will be analysed in section 5.4, as well as *López* (1994), *John et al.* (1995), *Saint-Paul* (1995), and *Ansuategi et al.* (1998) where we have only considered papers dealing with the EKC. *Keeler et al.* (1971) come closest to our assumptions: They retain $U_{CC} < 0$ and $U_{PP} < 0$ (which are not needed in static models), but use the normal good assumption for U_{CP}.

[4]For a derivation of this equation see appendix to chapter 5 on page 138.

This is simpler than[5]

$$\frac{dQ}{dY} = \frac{\pi U_{CC} + U_{CP}}{\pi^2 U_{CC} + 2\pi U_{CP} + U_{PP}} \tag{4.4}$$

which we would obtain using second derivatives of the utility function. The second advantage of using MRS_C and MRS_P follows immediately: Considering (4.4) and assuming as usual $U_{CC} < 0$ and $U_{PP} < 0$ while realizing that the sign of U_{CP} is ambiguous, we might conclude – and *McConnell* (1997, 391) actually drew this conclusion – that (4.4) can be negative if U_{CP} is positive and sufficiently large. However, looking at (4.3) we immediately see that dQ/dY is always positive (unless either environmental quality or consumption is not a normal good). Hence, in general, using MRS_C and MRS_P instead of second derivatives of the utility function prevents us from drawing any conclusions which violate the normal good assumption. Such conclusions could also be prevented by assuming $U_{CP} \leq 0$ along with $U_{CC} < 0$ and $U_{PP} < 0$ such that (4.1) and (4.2) are satisfied. These assumptions, however, are stronger than necessary. In general, since the equations are shorter when using our method, the calculations are more straightforward and the results can be interpreted more easily.

In chapter 5 we will use this method to reconsider *McConnell's* (1997) model in a more general framework. In chapter 8 we will show that we can also apply this method to an overlapping generations model adapted from *John* and *Pecchenino* (1994) because each generation solves a static problem (where the solution determines the initial condition of the next generation).[6]

---

[5]Inserting (4.1) and (4.2) into (4.3) we obtain $\{-U_{CC}U_P + U_{PC}U_C\}/\{-\pi U_{CC}U_P + \pi U_{PC}U_C - U_{CP}U_P + U_{PP}U_C\}$. Dividing through by U_C and using $\pi = -U_P/U_C$ (from the appendix to chapter 5) yields (4.4). Note that (4.4) is identical with equation (3) in *McConnell* (1997, 391).

[6]Similarly, in the model of *López* (1994) the term $a = -U_{CC}C/U_C$ (see (3.1) on page 65) has been derived assuming $U_{CP} = 0$. Using our method we find $a = -MRS_C C/MRS$. So *López* (1994) could have derived his results under less restrictive assumptions.

4.3 The elasticity of intertemporal substitution

When we leave the static framework and consider a dynamic model, matters become more involved. In a dynamic framework utility depends not only on current consumption (and pollution), but also on the consumption of all future periods. The utility function is of the form $\bar{U}(\cdot_0, \cdot_1, \cdot_2, \cdots)$ where the index denotes the period and where it is irrelevant on which variable(s) the utility function $\bar{U}$ depends. It is 'by far the most widely used' (*Deaton* 1992, 4) and plausible (*Stern* 1977) assumption that the utility function is additively separable, i.e. it is assumed that $\bar{U}$ can be written as

$$\bar{U} = \sum_{t=0}^{T}(1+\rho)^{-t}U(\cdot_t) \qquad \text{or} \qquad \bar{U} = \int_0^T e^{-\rho t}U(\cdot_t)dt$$

where $T \leq \infty$, t is time, $\rho \geq 0$ is the time preference rate, and where the function $U(\cdot)$ is called the felicity function (or the instantaneous utility function).

The assumption that utility is additively separable implies that we use a cardinal felicity function: We are comparing felicity *levels* at different points in time. Results are only robust to linear monotonic transformations of the felicity function (*Koopmans* 1965, 227 and *Barro* and *Sala-i-Martin* 1995, 61) because a linear monotonic transformation ($a + bU$ with $b > 0$) of the felicity function U is also a linear monotonic transformation ($a(1 - e^{\rho T})/\rho + b\bar{U}$) of the utility function $\bar{U}$ as

$$\int_0^T e^{-\rho t}(a + bU(\cdot_t))dt = \frac{a}{\rho}(1 - e^{\rho T}) + b\int_0^T e^{-\rho t}U(\cdot_t)dt.$$

A nonlinear monotonic transformation of the felicity function U, however, is an ordinal transformation of the utility function $\bar{U}$. Such a nonlinear monotonic transformation of the felicity function is a change in the elasticity of intertemporal substitution (*Stern*

1977). Perhaps the easiest way to see this is a simple example: For the felicity function

$$U(C) = \frac{C^{1-\sigma}}{1-\sigma} \tag{4.5}$$

where $\sigma > 0$ the elasticity of intertemporal substitution is constant and given by $1/\sigma$ (*Barro* and *Sala-i-Martin* 1995, 64). Applying the nonlinear monotonic transformation

$$\tilde{U} = \frac{(1-\sigma)^{\frac{1-\theta}{1-\sigma}}}{1-\theta} U^{\frac{1-\theta}{1-\sigma}} \tag{4.6}$$

where $\theta > 0$ and $\theta \neq \sigma$ leads to $\tilde{U}(C) = C^{1-\theta}/(1-\theta)$. Thus the constant elasticity of intertemporal substitution has changed to $1/\theta$. In general, at different levels of the variable(s) entering the felicity function attained at different points in time a nonlinear monotonic transformation of the felicity function changes the relative marginal effect on felicity of a change of these variable(s). For instance, suppose that due to a nonlinear monotonic transformation the positive marginal effect on felicity of an increase of the variable(s) entering the felicity function rises in the future relative to the present. Then the transformation implies that we reduce these variable(s) now to increase them later on, i.e. that we prefer faster growth. In fact, we will find in chapters 6 and 7 that a nonlinear monotonic transformation of the felicity function changes the growth path of the economy.

Henceforth, when we speak about monotonic transformations we always mean nonlinear ones. Furthermore, from now on we use the terms 'monotonic transformations of the felicity function' and 'changes in the elasticity of intertemporal substitution' interchangably.

It is interesting to analyse whether or not the results of a dynamic model are robust to changes in the elasticity of intertemporal substitution or in other words whether or not they are robust to monotonic transformations of the felicity function. As mentioned, we need a cardinal felicity function to compare felicity levels at different points in time. Therefore at least part of the results are not robust in each and every dynamic model. However, it turns out that in some models all crucial results are robust. The model of chapter 6 is an example. In other models as in the model of chapter 7, in contrast, the main result changes with a monotonic transformation of the felicity function. Hence, we need a method to analyse whether or not the results are robust. We present such a method in the next subsection.

4.3.1 A method to check for robustness

Results that only depend on the ordinal felicity function are robust to monotonic transformations of the felicity function because monotonic transformations only change the cardinal, not the ordinal felicity function. If only one variable, for example consumption C, enters the felicity function, i.e. $U = U(C)$, then the ordinal felicity functions just says that more is better. Thus all results which depend on the derivatives of U (U_C, U_{CC}, $U_{CCC}, \ldots$) change with monotonic transformations of the felicity function except for those results which depend only on the sign of U_C. If the felicity function is given by (4.5), *Barro* and *Sala-i-Martin* (1995, 76) analyse what happens when the constant elasticity of intertemporal substitution $1/\sigma$ changes, i.e. when we transform (4.5) according to (4.6). Their results are similar to those we will derive in chapters 6 and 7.

In what follows we develop a method how to check for robustness to changes in the elasticity of intertemporal substitution when two variables enter the felicity function. As an example we again use consumption and pollution. In fact, we can use a variation of our method to analyse static models. Thus the method can also be applied if more than two variables enter the felicity function. To my knowledge there are no attempts in the literature to check whether or not the results are robust to monotonic transformations of the felicity function if two or more variables enter the felicity function.

As mentioned results depending on the ordinal felicity function only are robust. Hence, results are robust if they only rely on the slope of the indifference curve at each point in figure 4.1, i.e. if they only depend on MRS, MRS_C, MRS_P, and higher order derivatives of MRS. In contrast, they are not robust if they *additionally* depend on second (U_{CC}, U_{PP}, U_{CP}) or higher order derivatives of U, or on the exact value (not only the sign) of U_C or U_P. However, MRS, MRS_C, and MRS_P can be written in terms of U_{CC}, U_{PP}, U_{CP}, U_C, and U_P (see (4.1) and (4.2)). Therefore we must be careful since results which seem to depend on derivatives of U, might possibly be rewritten in terms of MRS, MRS_C, and MRS_P. Note that we must allow for $U_{CP} \neq 0$: If we assumed $U_{CP} = 0$, we could not check whether or not the results are robust to nonlinear monotonic transformations of the felicity function.[7] In the literature, however, $U_{CP} = 0$ is often assumed to simplify the calculations. Then we cannot check whether or not the results are robust to changes of the elasticity of intertemporal substitution.

In chapters 6 and 7 we will use this method, i.e. we will analyse which results depend only on the MRS and its derivatives. In chapter 6 we examine a dynamic version of the static model of chapter 5 and in chapter 7 we reconsider the often cited model of *Selden* and *Song* (1995). These two models are almost identical: The only difference between the two models – after dropping two restrictive assumptions in *Selden* and *Song's* (1995) model – is that in chapter 6 pollution is generated by consumption, whereas in chapter 7 it is generated by production capital. This small difference has a huge effect: Almost all results of the model with pollution generated by consumption are robust to changes in the

[7] Consider the general monotonic transformation $\tilde{U} = f(U)$ where f is a strictly increasing function. Then $\tilde{U}_{CP} = f_{UU}U_C U_P + f_U U_{CP}$. So for $U_{CP} = 0$ we only find $\tilde{U}_{CP} = 0$ if $f_{UU} = 0$, i.e. if f is linear.

elasticity of intertemporal substitution. In contrast, in the model with pollution generated by capital the main result, i.e. whether the PIR is monotonically rising or inverted-U shaped, depends on the elasticity of intertemporal substitution: In the model of *Selden* and *Song* (1995) an EKC exists only if the elasticity of intertemporal substitution is sufficiently high.

4.4 Conclusion

In many static models there are assumptions about the sign of the second derivatives of the utility function. However, these are cardinal assumptions which are stronger than necessary. We propose to use only ordinal assumptions about the utility function, i.e. only assumptions about the slope of the indifference curves (i.e. the marginal rate of substitution) and about the change of this slope when we increase one of the variables entering the utility function. Hence, these assumptions have a straightforward graphical interpretation. They also ensure that all goods are normal goods. Therefore our method prevents us from drawing conclusions which violate the normal good assumption. This, however, can happen when we make the standard assumptions about the signs of the second derivatives of the utility function. Furthermore, our method allows us to write equations in a simpler, more compact way which facilitates the calculations and enables us to interpret the results more easily. Hence, we argue that we should not make assumptions about the sign of the second derivatives of the utility function. Instead, we should always use the normal good assumption. Thus static models can be solved with an ordinal utility function only.

In contrast, when we consider a dynamic model and when we assume as usual that utility is additively separable, we must use a cardinal felicity function since we must compare felicity levels at different points in time. Thus a nonlinear monotonic transformation of the felicity function is an ordinal change of the utility function. In fact, a nonlinear monotonic transformation is a change in the elasticity of intertemporal substitution. It is interesting to analyse which results are robust to changes in the elasticity of intertemporal substitution. We offer a simple method how this can be done. Results are robust if they depend on the ordinal felicity function only. Hence, they are robust if they rely only on the marginal rate of substitution and its derivatives. Since a cardinal felicity function must be used, at least part of the results are not robust in each and every dynamic model. However, we will show in chapter 6 that there are models in which all important results are robust to changes in the elasticity of intertemporal substitution.

5. A Simple Static Model: Reconsidering the Model of McConnell

5.1 Introduction [1]

After having considered two methodological issues in chapter 4, we now return to the analysis of the EKC. In the rest of part II of this book we will examine one of the empirically validated causes of the EKC: Rising demand for environmental quality. First, we will reconsider the simple static model of *McConnell* (1997) in this chapter. We will show that at low income levels no abatement is optimal and pollution increases with income. Once abatement expenditures are positive, we demonstrate that satiation in consumption is not only sufficient to find an EKC, but a tendency to satiation is even necessary if we assume a standard functional form for the pollution function. In chapter 6 we endogenize growth and thus make the model dynamic. We find that all the results of chapter 5 also apply in the dynamic framework. Finally, in chapter 7 we study the same model as in chapter 6, but we assume that pollution is generated by production capital instead of by consumption. This small change has a huge effect: In the model of chapter 7 the elasticity of intertemporal substitution is an important determinant of the shape of the PIR, i.e. whether it is monotonically rising, an EKC, or N-shaped.

We have shown in section 3.2 that the downturn of the EKC might emerge because demand for environmental quality increases with income. As mentioned in section 3.2 *Ansuategi et al.* (1998),

---

[1] This chapter has been published previously (Lieb (2002): The Environmental Kuzents curve and satiation: a simple static model. *Environment and Development Economics* 7, 429–448).

Neumayer (1998), *Vogel* (1999, 76), *Cole* (2000b, 27), and *de Bruyn* (2000, 89) all argue that environmental quality is a luxury good, i.e. that the income elasticity of demand for environmental quality is greater than one. However, *McConnell* (1997, 394) shows that 'there is no special significance of an income elasticity equal to one' for the shape of the PIR. Nevertheless, *McConnell* (1997, 394) claims that 'higher income elasticities [of demand for environmental quality] result in slower increases or faster declines in pollutants'.[2] We show, however, that this is not correct. Instead the relationship between the income elasticity and the PIR is ambiguous.

Reconsidering *McConnell's* (1997) model we find two other explanations why demand for environmental quality increases with income and thus why there is an EKC: First, because environmental quality is a normal good. Hence, what really matters is not that the income elasticity of demand for environmental quality is greater than one (i.e. that environmental quality is a luxury good), but that it is greater than zero (i.e. that environmental quality is a normal good). As we will demonstrate this can also be called a tendency to satiation in consumption, i.e. a tendency to zero marginal utility of consumption at a high, possibly infinite consumption level. There exist other models in the literature which show that a tendency to satiation is one possible cause of the EKC although the words 'satiation' and 'normal good' have never been used (see *López* 1994, *John* and *Pecchenino* 1994, *Selden* and *Song* 1995, and *Stokey* 1998). The second cause of the EKC is an increasing marginal rate of transformation between pollution and consumption. This condition, however, is not satisfied for standard functional forms for the pollution function. Hence, we contribute to the literature by showing that in a representative consumer model with pollution generated by consumption the tendency to satiation (i.e. the normality of environmental quality)

[2] A similar result is found by *Magnani* (2000a). But she uses a simplistic utility function which is linear in consumption and environmental quality.

is even *necessary* to find an EKC (assuming standard functional forms for the pollution function). Not surprisingly, outright satiation at a finite consumption level is always sufficient to generate an EKC. The intuition is that approaching the satiation level we can easily dispose of some consumption almost without lowering utility. Therefore additional income is better used for abatement such that pollution falls.

To derive our results we study a simple static model adapted from *McConnell* (1997). We deviate from and add to his work in several respects: First, while *McConnell* (1997) uses a cardinal utility function, we apply only an ordinal one: We employ the normal good assumption and the method developed in section 4.2. Second, we restrict pollution to be nonnegative. It turns out that this is important because we will show that in the model of *McConnell* (1997) pollution actually becomes zero at some high income level. Third, we assume that abatement is more efficient when emissions are high or, in other words, we do not require that the cross-derivative of the pollution function is zero. This enables us to derive two conditions on the pollution function which are of prime importance for the behaviour of pollution: Pollution does no longer have to be zero at high income levels. Fourth, *McConnell* (1997) only considers the interior solution with positive abatement expenditures. However, to generate an EKC it is crucial to study the possibility of zero abatement expenditures as well. Fifth, we offer a graphical interpretation of the results. Last, but not least, we draw different conclusions from the model.

We also prove that the model of *Stokey* (1998) – which at first glance seems rather different from our model – is actually an interesting special case of it. Her model is an ideal example to illustrate our findings. Moreover, we offer a new interpretation of *Stokey's* (1998) result: With asymptotic satiation there is an EKC, but without satiation pollution is monotonically rising. Thus in her model the shape of the PIR exactly depends on whether or not there is satiation.

Satiation is not a very popular concept with economists although there is some evidence in favour of it: A tendency to satiation for agricultural and industrial goods was already observed in the 1950s in the richer countries (*Fourastié* 1954, 86). *Zinn* (1987, 277) even claims to have seen evidence of satiation for certain services. Furthermore, a tendency to satiation is equivalent to the normality of environmental quality. That environmental quality is a normal good, however, is an assumption that most economists would subscribe to.

In the model analysed in this chapter we abstract from several empirically validated explanations of the EKC discussed in chapter 3: An improvement of political institutions which would allow a better internalization of external costs is not possible as we look at the first-best solution. A pollutant cannot be substituted by another one because there is only one pollutant in the model. Moreover, technological progress is excluded from the model. Furthermore, migration of dirty industries cannot be analysed because there is only one country. Hence, the tendency to satiation or the normality of environmental quality is only a necessary condition for an EKC if we abstract from these possibilities.

We must also be cautious because the model is only valid if the consumer – or the benevolent government – is aware of the causal links between consumption and pollution which in reality is not always the case (*Stigliani et al.* 1991, *Opschoor* 2000, 364, and *Huesemann* 2001).

This chapter is organized as follows. In section 5.2 we introduce the basic assumptions which are also used in chapters 6 and 7. The model is solved in section 5.3 where we analyse under what conditions an EKC emerges. Then the results are compared to the literature in section 5.4: We comment on the models of *McConnell* (1997) and *Stokey* (1998). Section 5.5 concludes.

5.2 Basic assumptions

The simple static model of *McConnell* (1997) is used and extended in this chapter. We assume a representative consumer with a given utility function U which depends on consumption C and pollution P, i.e. $U = U(C, P)$. We assume $U_C > 0$ (below the possible satiation level C^*) and $U_P < 0$ such that the *marginal rate of substitution* is $MRS = -U_C/U_P > 0$. Since the model is static, we apply the method developed in section 4.2, i.e. we assume that consumption C and environmental quality or $-P$ are normal goods: Empirical evidence that environmental improvements are normal goods can be found in *Kriström* and *Riera* (1996). As shown in (4.1) and (4.2) these assumptions can be written as

$$
\begin{aligned}
MRS_C &= \frac{-U_{CC}U_P + U_{PC}U_C}{U_P^2} < 0 \\
MRS_P &= \frac{-U_{CP}U_P + U_{PP}U_C}{U_P^2} < 0.
\end{aligned}
$$

Finally, we assume $\lim_{C \to 0} MRS = \infty$ for all P. Hence, we make only ordinal assumptions about the utility function, while *McConnell* (1997) made assumptions about the sign of the second derivatives of the utility function. As shown in section 4.2 these assumptions are cardinal and less general than necessary. Note finally that if environmental quality is a normal good, MRS_C is negative. Then the slope of the indifference curve becomes increasingly flatter when we move to the right in figure 4.1 on page 104. Therefore we could also call this a tendency to satiation in consumption because at the satiation level the indifference curve becomes horizontal. Hence, we will use the terms 'normality of environmental quality' and 'tendency to satiation' interchangeably.

Next, pollution P depends on consumption C and abatement expenditures A (in terms of the consumption good), i.e. $P = P(C, A)$. We assume that $P_C > 0$, $P_{CC} \geq 0$, $P_A < 0$, $P_{AA} > 0$,

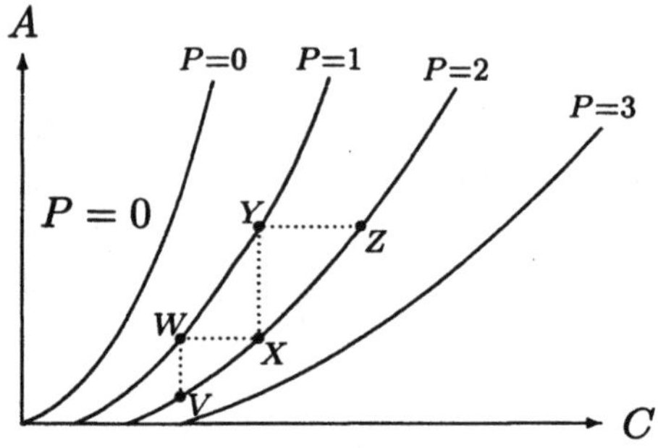

Figure 5.1: The iso-pollution lines.

$P_{AC} = P_{CA} \leq 0$, and $\lim_{A \to 0} |P_A| < \infty$.[3] So we allow for the case that consumption causes emissions linearly ($P_{CC} = 0$). Abatement expenditures become increasingly less efficient as the cheapest abatement opportunities are first exploited ($P_{AA} > 0$, for empirical evidence see *Faber et al.* 1996, 272). Contrary to the bulk of literature which for simplicity assumes $P_{AC} = 0$,[4] we assume $P_{AC} = P_{CA} \leq 0$ which is more general and plausible: With more emissions caused by consumption the abatement expenditures are more efficient (P_{AC}) or with higher abatement expenditures the polluting effect of more consumption is smaller (P_{CA}).[5]

[3]Although we assume $P_C > 0$, $P_A < 0$, and $P_{AA} > 0$, we do not exclude the possibility that $\lim_{C+A \to \infty} P_C = 0$, $\lim_{C+A \to \infty} P_A = 0$, and $\lim_{C+A \to \infty} P_{AA} = 0$ where $C + A$ is optimally allocated between C and A.

[4]This assumption can be found in *McConnell* (1997), as well as in *Forster* (1973), *Gruver* (1976), *John* and *Pecchenino* (1994), *John et al.* (1995), *Selden* and *Song* (1995), *Ansuategi et al.* (1998), *Ansuategi* and *Perrings* (2000), and *Ansuategi* (2000, chapter 4).

[5]Suppose for example that a certain abatement technology can abate a certain percentage of emissions. A higher percentage is achieved by a more expensive technology. Then the higher consumption, i.e. the higher emissions, the higher are the emissions which can be abated by a given technology (P_{AC}). Similarly, the higher abatement expenditures, the less polluting is higher consumption because a higher percentage of the additional emissions is abated (P_{CA}).

To derive our results it is convenient to introduce two further conditions. To visualize these conditions consider the iso-pollution lines in figure 5.1.[6] Since pollution must always be nonnegative, there are no iso-pollution lines above the $P = 0$-line. From $dP = 0 = P_C dC + P_A dA$ we see that the slope of the iso-pollution lines is $dA/dC = -P_C/P_A > 0$. The two conditions are

$$\begin{aligned} V := \left.\frac{dP_C}{dC}\right|_{P=\text{constant}} &= P_{CC} + P_{CA} \left.\frac{dA}{dC}\right|_{P=\text{constant}} \\ &= \frac{-P_{CC}P_A + P_{AC}P_C}{-P_A} < 0 \qquad (5.1) \\ W := \left.\frac{dP_A}{dC}\right|_{P=\text{constant}} &= P_{AC} + P_{AA} \left.\frac{dA}{dC}\right|_{P=\text{constant}} \\ &= \frac{-P_{AC}P_A + P_{AA}P_C}{-P_A} > 0. \qquad (5.2) \end{aligned}$$

The first condition states that P_C falls when we move along the iso-pollution lines to the right. Since the horizontal distance between two iso-pollution lines is determined by P_C, (5.1) also means that the horizontal distance increases with consumption ($\overline{YZ} > \overline{WX}$ in figure 5.1). In other words, consumption becomes less polluting with rising income if we allocate additional income to consumption and abatement such that pollution stays constant. Condition (5.2) states that $|P_A|$ falls when we move along the iso-pollution lines to the right. Since the vertical distance between two iso-pollution lines is determined by P_A, (5.2) also means that the vertical distance increases with abatement ($\overline{XY} > \overline{VW}$). Put differently, abatement becomes more expensive with rising income if pollution is again held constant. Both conditions might or might not hold. In appendix A on page 259 it is shown that two standard functional forms fulfil conditions (5.1) and (5.2). If we assumed $P_{AC} = 0$, as is common in the literature, we would implicitly

[6]The iso-pollution lines cannot only be convex as in figure 5.1 but they might also be concave.

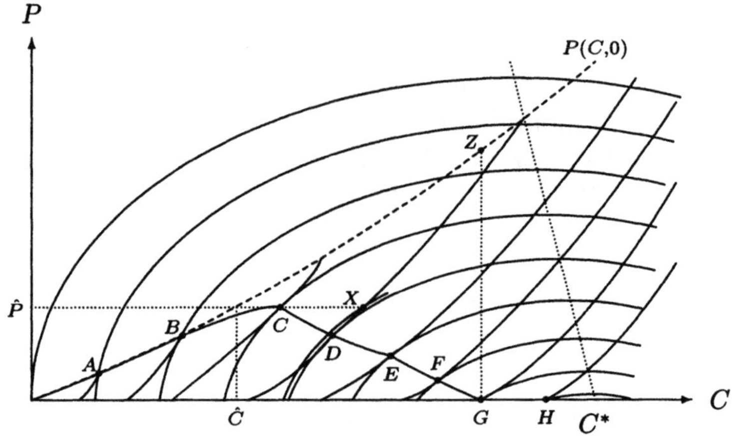

Figure 5.2: Indifference curves, CPCs, and the optimal path of the economy.

assume that (5.2) is satisfied, but (5.1) is not. Note that all the assumptions made so far will also be used in chapters 6 and 7.

To complete the static model, we assume that the exogenously given income Y is split up into consumption C and abatement expenditures A, i.e. $Y = C + A$.

5.3 Solution of the model

The representative consumer or the social planer maximizes

$$\max_{C,A} U(C, P(C, A)) + \lambda(Y - C - A) + \gamma A + \mu P(C, A),$$

where λ, γ, and μ are the multipliers with respect to $Y = C + A$ and the nonnegativity constraints $A \geq 0$ and $P \geq 0$, respectively.[7]

[7] *McConnell* (1997) assumes an interior solution for A and wholly neglects that $P \geq 0$ must hold.

The first order conditions are

$$U_C + (U_P + \mu)P_C - \lambda = 0 \tag{5.3}$$
$$(U_P + \mu)P_A - \lambda + \gamma = 0 \tag{5.4}$$
$$Y - C - A = 0. \tag{5.5}$$

The *consumption possibilities curve* (CPC) shows all feasible combinations of consumption and pollution for a given income level Y: It is defined by $P(C, Y - C)$ and is depicted in figure 5.2. The slope of the CPC is the *marginal rate of transformation* (MRT) between pollution and consumption, i.e. $MRT = P_C - P_A > 0$. The CPC is convex as $P_{CC} - 2P_{AC} + P_{AA} > 0$. If income increases, the CPC moves downwards since more abatement is possible for the same level of consumption. The endpoints of the $CPCs$ lie on the convex $P(C,0)$-line (points above the $P(C,0)$-line would imply negative abatement expenditures and are thus not feasible). Combining (5.3) and (5.4) and dividing by $-U_P$ we derive

$$MRS = -\frac{U_C}{U_P} = (P_C - P_A)\frac{U_P + \mu}{U_P} - \frac{\gamma}{U_P} = MRT\frac{U_P + \mu}{U_P} - \frac{\gamma}{U_P} \tag{5.6}$$

or $MRS = MRT$ at an interior solution where $\mu = \gamma = 0$. Hence, at an interior solution, as at points C, D, E, and F in figure 5.2, the concave indifference curve is tangent to the convex CPC.

5.3.1 Zero abatement expenditures

If on the $P(C,0)$-line the slope of the indifference curve is steeper than the slope of the CPC as at point A in figure 5.2, then the point of the CPC on the $P(C,0)$-line is optimal. Thus abatement expenditures are zero. In other words, if on the $P(C,0)$-line

$$MRS(C, P(C,0)) \geq MRT(C,0) = P_C(C,0) - P_A(C,0) \tag{5.7}$$

holds, (5.6) shows that $A = 0$ since $\gamma \geq 0$ (where $P > 0$ ($\mu = 0$) on the $P(C,0)$-line).

Proposition 5.1 *There exists an income level $\hat{Y}$ such that at low income levels ($Y < \hat{Y}$) abatement expenditures are zero ($A = 0$) and pollution increases with exogenously growing income ($dP/dY > 0$). At a higher level of income ($Y = \hat{Y}$), however, abatement expenditures become positive.*

Proof. Since $\lim_{C\to 0} MRS = \infty$ whereas $(P_C - P_A)$ is bounded for small Y ($\lim_{A\to 0} |P_A| < \infty$), (5.7) holds for small Y. Hence, if Y rises and $A = \mu = 0$, pollution grows ($dP/dY = P_C dC/dY + P_A dA/dY = P_C > 0$ as $dC/dY = 1$ and $dA/dY = 0$), and (5.6) changes according to

$$MRS_C + MRS_P P_C = P_{CC} - P_{AC} - d(\gamma/U_P)/dY.$$

Because the terms on the left hand side are negative (see (4.1) and (4.2)) and the first two terms on the right hand side are positive, $-d(\gamma/U_P)/dY$ must be negative or $|\gamma/U_P|$ must decline. As soon as $\gamma = 0 = \gamma/U_P$ which happens at income level $\hat{Y}$ (that is the definition of $\hat{Y}$), abatement expenditures become positive (at point B in figure 5.2). □

At low income levels the economy is so poor or suffers from so little pollution that it is optimal not to abate. If income then rises, all additional income is spent on consumption as long as abatement is zero. Then pollution rises along the $P(C, 0)$-line. Abatement expenditures become positive at a low income level if the indifference curves are flat and the $CPCs$ are steep (see (5.7)). The MRS is small if consumption is relatively satiated (U_C small) or if preferences are 'green' ($|U_P|$ high): A rise in pollution decreases utility by a greater amount if the environment is highly valued. This finding also explains why local pollutants or pollutants with severe health effects (for which $|U_P|$ is high) are abated at lower income levels than transboundary and global pollutants or less health damaging pollutants (see *Ansuategi* and *Perrings* 2000). Abatement expenditure also become positive at a small income level, if they can be abated at low costs ($|P_A|$ high, for example CFCs, see *Cole et al.* 1997) or if consumption is highly pollution-intensive (P_C high).

5.3.2 The behaviour of pollution at an interior solution

Let us analyse the behaviour of pollution once abatement expenditures are positive.

Proposition 5.2 *At an interior solution consumption and abatement expenditures increase with exogenously rising income, and pollution declines with income if*

$$-MRS_C > \frac{P_{CC}P_A - P_{AC}P_C - P_{AC}P_A + P_{AA}P_C}{-P_A} = W - V, \tag{5.8}$$

where V and W are defined in (5.1) and (5.2). Pollution rises if the inequality sign in (5.8) is reversed.

Proof. Totally differentiating (5.6) and (5.5) at an interior solution with positive abatement expenditures and positive pollution ($\gamma = \mu = 0$) yields

$$\begin{bmatrix} MRS_C + MRS_P P_C - P_{CC} + P_{AC} & MRS_P P_A - P_{AC} + P_{AA} \\ -1 & -1 \end{bmatrix} \begin{bmatrix} dC \\ dA \end{bmatrix} = \begin{bmatrix} 0 \\ -dY \end{bmatrix}$$

Denote the entries in the first row of the 2×2 matrix by $b_1 < 0$ and $b_2 > 0$. The determinant of this matrix is $b_2 - b_1 > 0$. By Cramer's rule we obtain

$$\frac{dC}{dY} = \frac{b_2}{b_2 - b_1} > 0 \qquad \text{and} \qquad \frac{dA}{dY} = \frac{-b_1}{b_2 - b_1} > 0,$$

where $dC/dY + dA/dY = 1$. Next, using $dP/dY = P_C dC/dY + P_A dA/dY$ we derive[8]

$$\frac{dP}{dY} = \frac{P_{CC}P_A - P_{AC}P_C - P_{AC}P_A + P_{AA}P_C - MRS_C P_A}{P_{AA} - 2P_{AC} + P_{CC} + MRS_P(P_A - P_C) - MRS_C}, \quad (5.9)$$

from which (5.8) follows immediately. □

Rising income increases consumption and abatement expenditures as was to be expected since consumption and abated pollution are normal goods. Empirical evidence that abatement rises with income can be found in *Hettige et al.* (1996, 1899). To grasp the intuition behind (5.8), we derive it graphically, following figure 5.2. We start at an optimal point like C where the indifference curve is tangent to the CPC. Consider point X which lies on the new CPC (for the higher income) at the same level of pollution as before. Pollution falls (rises) with growing income if the indifference curve is flatter (steeper) than the CPC at point X. The new optimal point is D. The slope of the CPC, i.e. the MRT, changes on a horizontal ray on which pollution is constant according to $\left.\frac{d(P_C - P_A)}{dC}\right|_{P=constant} = V - W$. The slope of the indifference curve, i.e. the MRS, falls along a horizontal ray ($MRS_C < 0$, see (4.1)). Pollution declines if the MRS falls more than the MRT, i.e. if $-MRS_C > W - V$ which is (5.8). A corollary follows immediately:

Corollary 5.1 *Pollution falls with income at an interior solution if the MRT increases when we move to the right in figure 5.2, i.e. if $V - W \geq 0$. This is the case if conditions (5.1) and (5.2) are both violated.*

For example if $P_{AA} = P_{AC} = 0$, i.e. if there are constant returns to scale in abatement, we obtain $V > 0$ and $W = 0$. Thus pollution

[8]For $P_{AC} = 0$ (5.9) is identical with (4) in *McConnell* (1997, 391): Expanding (5.9) with $-U_P$ and inserting (4.1) we find $\{-U_P(P_C P_{AA} + P_A P_{CC}) - P_A U_{CC} + P_A U_{PC} U_C / U_P\} / \{-U_P(b_2 - b_1)\}$ where according to (5.6) at an interior solution U_C/U_P equals $P_A - P_C$.

falls with income.[9] However, as mentioned in section 5.2 there is empirical evidence for $P_{AA} > 0$. Furthermore, for standard functional forms (5.1) and (5.2) are both satisfied (see appendix A on page 259) such that $V - W < 0$. Then the slope of the indifference curve and the slope of the CPC are both smaller at point X than at point C in figure 5.2. In general, it follows from (5.8) that pollution falls

- if the slope of the indifference curves, i.e. the MRS, falls fast with consumption holding pollution constant ($|MRS_C|$ high). This is the case
  - if there is a tendency to satiation (U_C falls fast with consumption) and
  - if preferences quickly become 'greener' with rising consumption: Then starting at the same level of pollution, the marginal disutility of a rise in pollution, i.e. $|U_P|$, grows fast with consumption.
- if the slope of the CPC, i.e. the MRT, becomes slowly flatter (i.e. slowlier than the MRS) or even steepens when we move horizontally to the right in figure 5.2. This is the case
  - if abatement costs grow slowly or even shrink when we move along the iso-pollution lines to the right (see figure 5.1, W slightly positive or negative) and
  - if the environmental impact of consumption declines slowly or even rises with income holding pollution constant (V slightly negative or positive).

Thus the behaviour of pollution depends on the relationship between preferences and the pollution function (for an example see

[9]The fact that constant returns cause the EKC is also derived by *Selden* and *Song* (1995) and it resembles the result of *Andreoni* and *Levinson* (2001) which we discuss in section 5.4.

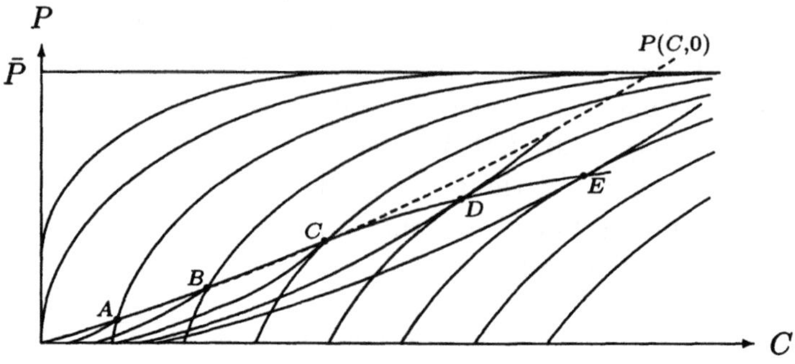

Figure 5.3: The path of the economy if $\lim_{P \to \bar{P}} U_P(C, P) = \infty$ *for all* C.

section 5.4). Since pollution rises as long as abatement expenditures are zero, we find an EKC if pollution falls at the interior solution.

It follows that if the MRT declines ($V - W < 0$) as it does for standard functional forms, the MRS must fall even faster to cause an EKC. Hence, a corollary ensues:

Corollary 5.2 *For standard functional forms for the pollution function a sufficiently strong tendency to satiation is necessary for an EKC to exist or in other words it is necessary that environmental quality is a normal good.*

Note that if there are disastrous environmental damages beyond a certain level of pollution $\bar{P}$, causing U_P to become $-\infty$, the indifference curves become horizontal at $\bar{P}$ as shown in figure 5.3. Hence, $\bar{P}$ will never be reached as the CPC always has a positive slope so that the tangential point will be below $\bar{P}$ for every level of income.

5.3.3 Zero pollution and satiation

Corollary 5.1 tells us that there is an EKC if the MRT increases ($V-W \geq 0$). From now on, however, we concentrate on a declining MRT ($V - W < 0$). To find weaker conditions for an EKC, we study under what conditions zero pollution is reached. The assumption that zero pollution is technically feasible will be relaxed in subsection 5.3.4: We will show that the results derived assuming that zero pollution is technically feasible hold also if zero pollution requires infinite abatement expenditures.

Pollution is zero whenever the slope of the indifference curve is smaller than the slope of the CPC on the C-axis in figure 5.2 (see points G and H). In other words, pollution is zero, i.e. $\mu \geq 0$, whenever

$$MRS(C,0) \leq MRT(C,A) = P_C(C,A) - P_A(C,A) \qquad (5.10)$$

holds at the point of the CPC where pollution is zero (see (5.6)).[10] If for every point of the C-axis there is a CPC starting at this point, we say that zero pollution is technically feasible. Then the point of the CPC where pollution is zero moves to higher consumption levels as income rises.

Proposition 5.3 *Pollution is zero at high income levels if zero pollution is technically feasible and if for $C^* \leq \infty$*

$$\lim_{C \to C^*} MRS(C,0) < \lim_{C \to C^*} MRT(C,A) \qquad (5.11)$$

holds at the point of the CPC where pollution is zero. (5.11) is satisfied if at least one of the following three conditions holds:

---

[10]Cancelling $-P_A$ (5.10) can also be written as $U_C \leq -U_P P_C$. Thus if the marginal utility of consumption is smaller than the marginal disutility of pollution caused by consumption, pollution is zero.

(a) *Condition (5.1) is violated and* $\lim_{C\to\infty} MRS(C,0) = 0$.

(b) *Condition (5.2) is violated and* $\lim_{C\to\infty} MRS(C,0) = 0$.

(c) *There is a finite satiation level of consumption.*

Proof. Since zero pollution is technically feasible, the point of the CPC on the C-axis moves to higher consumption levels when income rises. Therefore it follows immediately from (5.10) that pollution is zero at high income levels if (5.11) holds. (5.11) is satisfied if $\lim_{C\to\infty} MRS(C,0) = 0$ and if either P_C or $-P_A$ increases or stays constant, i.e. if either condition (5.1) or (5.2) is violated. If P_A became zero, the CPC would not move downwards any more with rising income and zero pollution would not be technically feasible. Thus $P_A < 0$ for finite A and C. Hence, at C^*, i.e. at the satiation level of consumption, $MRS(C^*,0) = 0$ while $MRT = P_C - P_A > 0$. Thus (5.11) holds. □

Corollary 5.3 *If pollution is zero at high income levels, the PIR is an EKC.*

Proof. This follows immediately from proposition 5.1: Pollution rises at low income levels but is zero at high income levels. □

However, if (5.1) and (5.2) are not both violated, it is conceivable (see proposition 5.2) that before turning down, the PIR continues to rise when A becomes positive and that there are rising parts in the falling branch of the EKC.

Let us examine the intuition behind condition (a), (b), and (c) of proposition 5.3 in turn. Assume that (a) or (b) holds. Then consumption becomes less attractive since the MRS falls and approaches zero, i.e. since there is a tendency to satiation or in other words since environmental quality is a normal good. Note that $\lim_{C\to\infty} MRS(C,0) = 0$ is the usual Inada condition. Furthermore, most of the additional income is devoted to abatement because the environmental impact of consumption is rising ((5.1) violated) or because abatement has a large effect since its costs are falling ((5.2) violated). Pollution accordingly falls all the way down to zero. If P_{AC} was zero as in the model of *McConnell* (1997), (5.1) would be violated and pollution would become zero (if this is feasible and if $\lim_{C\to\infty} MRS(C,0) = 0$). However, as shown in appendix A on page 259 standard functional forms fulfil conditions (5.1) and (5.2). So the result that satiation (condition (c)) causes zero pollution and thus induces an EKC is more important. Intuitively, at the satiation level we can easily dispose of some consumption almost without lowering utility. If the same income used for abatement instead of consumption can decrease pollution, it is better used for abatement. Decreasing pollution is no longer possible when pollution is zero. Thus the satiation level can only be reached if pollution is zero. Pollution already becomes zero before the satiation level is reached (at point G in figure 5.2). At the satiation level the shadow price of income λ is zero because at point $(C^*, 0)$ in figure 5.2 the maximal utility level is reached.[11]

Finally, consider (5.11). In general, if condition (5.1) and (5.2) hold and if there is no satiation, the slope of the indifference curves, i.e. the MRS, and the slope of the CPC, i.e. the MRT, both fall when we move along a horizontal ray to the right in figure 5.2, and anything goes (see (5.8) or figure 5.2): If the MRS falls slowlier than the MRT, pollution rises monotonically. If both slopes fall at the same speed, pollution becomes constant at a positive level of pollution (this happens for $\sigma = 1$ in *Stokey's* (1998) model discussed in section 5.4). Finally, if the MRS falls faster than the MRT, pollution is falling. Analysing the behaviour of the two slopes on the C-axis in figure 5.2, we see that pollution becomes zero (at $\bar{C}$ in figure 5.4a) if (5.11) holds.[12] However, if the

[11]Since $MRS = \gamma = 0$ at C^*, it follows from (5.6) that $MRT(U_P + \mu) = 0$. Therefore $\mu = -U_P > 0$ and pollution is zero. With $\gamma = U_P + \mu = 0$ it follows from (5.4) that $\lambda = 0$.

[12]Note that the MRS falls always faster than the MRT in figure 5.4a and 5.4b. Conditions (a), (b), and (c) of proposition 5.3 are violated in figure 5.4 since $\lim_{C\to\infty} MRS(C,0) > 0$ and if both (a) and (b) held, the MRT would rise.

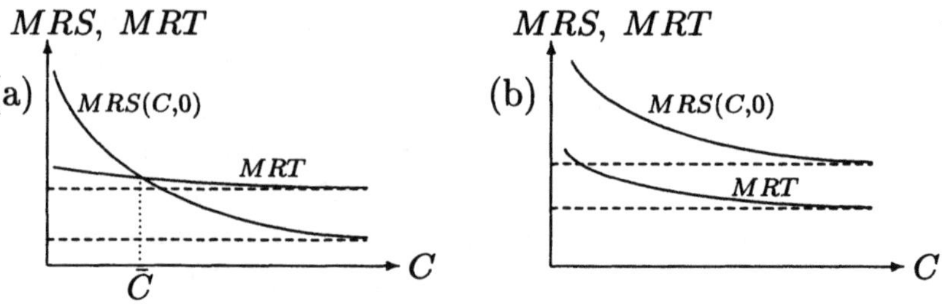

Figure 5.4: The MRS and the MRT at zero pollution if (a), (b), and (c) are violated.

inequality sign in (5.11) is reversed as in figure 5.4b, zero pollution is not attainable implying that the falling pollution asymptotically approaches some positive level from above. In the borderline case of $\lim_{C\to\infty} MRS = \lim_{C\to\infty} MRT$ the relative speed, with which the common limit is approached, determines whether or not zero pollution is attainable.

Therefore to bring about zero pollution, it is necessary and sufficient that the MRS becomes smaller than the MRT (see (5.11)). So we again find that *the normality of environmental quality (or a tendency to satiation: $MRS_C < 0$) is necessary for zero pollution and the EKC if the MRT declines along the C-axis in figure 5.2 as it does for standard functional forms.* Conditions (a), (b), and (c) in proposition 5.3 are sufficient for zero pollution, but not necessary. However, pollution becomes zero at a lower consumption level if one or several of the three conditions is satisfied: According to figure 5.4 zero pollution will be reached at a lower consumption level if the $MRT = P_C - P_A$ is rising at zero pollution ((a) and (b)) and if the $MRS(C, 0)$ is low ((c), assuming that satiation causes the MRS to be smaller everywhere). Remember that a low MRS and a high MRT (on the $P(C, 0)$-line) also let abatement expenditures become positive at a lower income level (see (5.7)).

5.3.4 The solution if zero pollution is technically infeasible

Up to now we have assumed that zero pollution is technically feasible. This need not be the case. It is often assumed that zero pollution requires infinite abatement expenditures. Then the $P = 0$-iso-pollution line in figure 5.1 is identical with the A-axis and all the $CPCs$ in figure 5.2 start at the origin. For rising income the CPC moves downwards but becomes identical with the C-axis only for infinite income. Hence, for all $C > 0$ zero pollution is technically not feasible since there is no CPC starting at point $(C, 0)$ in figure 5.2. However, for all positive levels of pollution – for example for $\hat{P} > 0$ in figure 5.2 – it is technically feasible to reach $\hat{P}$ which we define to mean that for all C[13] there is a CPC going through the point $(C, \hat{P})$. With rising income the CPC moves to the right on a horizontal line at $\hat{P}$. Note that we can choose $\hat{P}$ arbitrarily close to zero.[14] We can now state the analogue to proposition 5.3 for an interior solution.

Corollary 5.4 *Suppose that $\hat{P} > 0$ is technically feasible. Then if*

$$MRS(C, \hat{P}) < MRT(C, A) \tag{5.12}$$

holds at the point of the CPC where $P = \hat{P}$, pollution is smaller than $\hat{P}$. (5.12) is satisfied at high income levels if at least one of the following three conditions holds:

[13] More precisely, for all $C > \hat{C}$ where $\hat{C}$ is defined by $\hat{P} = P(\hat{C}, 0)$ and is shown in figure 5.2.

[14] Consider the two examples in appendix A on page 259, $P = C^{\beta}(C + A)^{1-\beta} - x$ where $\beta > 1$ and $P = fC^{g}/(kA + 1)^{h} - x$. In both examples zero pollution is technically not feasible if $x = 0$ (only for $A = \infty$ pollution becomes zero). However, in both examples an ever so small $\hat{P} > 0$ is technically feasible for every (finite) level of consumption if abatement is only high enough.

(a) *Condition (5.1) is violated and* $\lim_{C\to\infty} MRS(C,\hat{P}) = 0$.

(b) *Condition (5.2) is violated and* $\lim_{C\to\infty} MRS(C,\hat{P}) = 0$.

(c) *There is a finite satiation level of consumption.*

Proof. Decreasing consumption while holding income constant (which implies $dA = -dC$) we find that the left hand side of (5.12) increases ($MRS_C dC + MRS_P(P_C - P_A)dC > 0$ for $dC < 0$) and the right hand side decreases ($(P_{CC} - 2P_{AC} + P_{AA})dC < 0$) until $MRS = MRT$ which holds an interior solution (see (5.6)). Thus with less consumption and more abatement pollution is smaller. The proof that condition (a), (b), and (c) imply that (5.12) holds is exactly analogous to the proof of proposition 5.3. □

That pollution is smaller than $\hat{P}$ if (5.12) holds, can immediately be seen from point X in figure 5.2. As mentioned, if zero pollution requires infinite abatement expenditures, we can choose $\hat{P}$ arbitrarily close to zero. Therefore pollution approaches zero as consumption goes to infinity or to the satiation level. Hence, *we also find an EKC if zero pollution requires infinite abatement expenditures* (and if corollary 5.4 holds). Note that if (5.2) is violated, $|P_A|$ increases with income when pollution is held constant. Thus zero pollution is feasible in this case. If zero pollution is not feasible, (5.2) holds. If $P_{AC} = 0$ holds, (5.1) is violated. In this case we find an EKC (if $\hat{P}$ is feasible for all $\hat{P} > 0$ and if the Inada condition $\lim_{C\to\infty} MRS(C,\hat{P}) = 0$ holds). Hence, *it is crucial to allow for* $P_{AC} < 0$ *to make a monotonically rising PIR a possible outcome of the model.* Furthermore, the assumption that zero pollution requires infinite abatement expenditures makes only sense in a model with $P_{AC} < 0$. It again follows from (5.12) that if the MRT declines when we move to the right, the tendency to satiation must be strong enough to cause an EKC or in other words environmental quality must be a normal good.

5.4 Relationship to the literature

There are two other authors whose models are closely connected to the model of this chapter. The first is *McConnell* (1997). Although his intuition is similar, his interpretation is different: *McConnell* (1997, 391) claims that 'the higher the income elasticity of demand for environmental quality, the slower the growth in pollution when positive or the faster the decline in pollution when negative.' This, however, is not correct as shown by a counterexample in the appendix to this chapter. Rather the relationship between the income elasticity of demand for environmental quality and the PIR is ambiguous in our model.

The second author is *Stokey* (1998). Her model, seeming quite different at first glance, boils down to a special case of our general model with specific functional forms for $U(C, P)$ and $P(C, A)$. It is therefore an ideal example to illustrate our results. *Stokey* (1998) analyses the EKC in a model in which $C = zY$ where $z \in [0, 1]$ is an index for the technology used in production. A higher z allows more consumption, but also causes more pollution as $P = \max\{Yz^\beta - x, 0\}$, where $\beta > 1$ and $x \geq 0$ (*Stokey* uses $x = 0$ which makes zero pollution technically infeasible). Assuming $P > 0$ and using $z = C/Y$ we derive

$$P = C^\beta Y^{1-\beta} - x = C^\beta (C + A)^{1-\beta} - x.$$

which is a pollution function satisfying all our assumptions (see appendix A on page 259). Another formulation of the model is $C = (P + x)^{1/\beta} Y^{1-1/\beta}$, a Cobb-Douglas production function (*Stokey* 1998, 8). The utility function is given by[15]

$$U(C, P) = V(C) - H(P) = \frac{C^{1-\sigma} - 1}{1 - \sigma} - \frac{B(P + \epsilon)^\gamma}{\gamma}, \qquad (5.13)$$

[15] Despite *Stokey's* (1998, 7) stated wish to use a constant elasticity function $V(C)$ has no constant elasticity. Only $W(C) = C^{1-\sigma}/(1 - \sigma)$ has a constant elasticity ($W_C C/W = C^{-\sigma} C(1 - \sigma)/C^{1-\sigma} = 1 - \sigma$). The advantage of $V(C)$ is that we can use de l'Hospital's rule to show that $\lim_{\sigma \to 1} V(C) = \ln C$.

where $B, \sigma > 0$ and $\gamma > 1$. Note that we have extended *Stokey's* (1998) model by $\epsilon \geq 0$ to allow for zero pollution.[16] This utility function fulfils all our assumptions (see (4.1), (4.2), and $\lim_{C \to 0} MRS = \infty$).

At low levels of income we again find that pollution rises with income as long as there is no abatement. For an interior solution we derive:

Corollary 5.5 *In* Stokey's *(1998) model (5.8) is equivalent to* $\sigma > 1$. *Thus – at an interior solution – pollution falls with income if* $\sigma > 1$, *and it rises if* $\sigma < 1$.

Proof. Remember that (5.8) was $-MRS_C > [P_{CC}P_A - P_{AC}P_C - P_{AC}P_A + P_{AA}P_C]/(-P_A)$. Inserting (5.13) into (4.1) and dividing $(A.4) - (A.3)$ by $-(A.2)$ (all taken from appendix A on page 259) we find that (5.8) simplifies to

$$\frac{\sigma C^{-\sigma-1}}{B(P+\epsilon)^{\gamma-1}} > \beta C^{\beta-2} Y^{1-\beta}. \tag{5.14}$$

From (5.6) we know that $U_C = -U_P(P_C - P_A)$ at an interior solution. So inserting (5.13) and $(A.1) - (A.2)$ from appendix A we obtain

$$C^{-\sigma} = B(P+\epsilon)^{\gamma-1}\beta C^{\beta-1} Y^{1-\beta} \quad \text{or} \quad \frac{C^{-\sigma-1}}{B(P+\epsilon)^{\gamma-1}} = \beta C^{\beta-2} Y^{1-\beta}.$$

Inserting this result into (5.14) we immediately see that it reduces to $\sigma > 1$. □

Since in *Stokey's* model condition (5.1) and (5.2) hold (see (A.3) and (A.4) in appendix A), $\sigma \gtrless 1$ determines whether or

[16]For $\epsilon = 0$ we have $U_P(C, 0) = 0$ and $MRS(C, 0) = \infty$ for all C. Then (5.10) is never satisfied except at the satiation level where $0 = U_C(C^*, 0) = -U_P(C^*, 0)MRT = 0$. So if $U_P(C, 0) = 0$, only satiation causes zero pollution. In the literature utility functions with $U_P(C, 0) = 0$ are often used to simplify the calculations. At least for high consumption levels, however, it is not implausible that the marginal disutility at zero pollution ($U_P(C, 0)$) is negative: The first speck of pollution does diminish our utility.

not the MRS falls faster than the MRT along a horizontal ray in figure 5.2 at an interior solution (see (5.8)). As seen from the proof both the assumptions about $U(P, C)$ and $P(C, A)$ are crucial for obtaining an EKC if the elasticity of marginal utility of consumption ($-V_{CC}C/V_C = \sigma C^{-\sigma-1}C/C^{-\sigma} = \sigma$) is greater than one. *Stokey* (1998, 3 and 23) stops with this finding. However, looking at the definition of $V(C)$ in (5.13) we realize the following: On the one hand, if $\sigma < 1$, $V(C)$ is positive, and as C goes to infinity, $V(C)$ goes to infinity too. On the other hand, if $\sigma > 1$, $V(C)$ is negative, and as C goes to infinity, $V(C)$ asymptotically approaches $1/(\sigma - 1)$ from below. This means that utility of consumption is bounded above by $1/(\sigma - 1)$. Thus there is an EKC if utility of consumption is bounded above because then further consumption becomes less attractive. If utility of consumption is unbounded, however, pollution increases monotonically as consumption is more appealing.[17] This result is very intuitive. Note that assuming utility of consumption to be bounded above rests on cardinal utility. Thus this result cannot be generalized to ordinal utility. However, it is very similar to satiation. Therefore we conclude:

Corollary 5.6 *If there is asymptotic satiation in* Stokey's *(1998) model, the PIR is an EKC. If there is no satiation, the PIR is monotonically rising.*

In *Stokey's* (1998) model the satiation level C^* is ∞. So even when the consumption level is still much smaller than C^*, the tendency to satiation is so strong if $\sigma > 1$ that pollution declines already at very low levels of consumption (as soon as $A > 0$).

In a similar model as presented in section 5.3, *Andreoni* and *Levinson* (2001) find yet another reason for pollution to become zero at high levels of income (cf. section 3.5): They assume $P =$

[17]For $\sigma_1 > 1 > \sigma_2$ and for every point with $C > 1$ in figure 5.2 the $MRS = -U_C/U_P$ is higher for the smaller σ_2 because $C^{-\sigma_2} > C^{-\sigma_1}$.

$\alpha C - R(C, A)$ where $\alpha > 0$.[18] Here αC denote emissions and $R(C, A)$ is the reduction in emissions due to abatement where $R(C, A)$ is an increasing, concave function with increasing returns to scale and $R(C, 0) = R(0, A) = 0$. Then pollution becomes zero at some income level due to the increasing returns in abatement if $U_P(C, 0) < 0$. However, this income level may be high indeed as the proof rests on a limit argument when income goes to infinity.

5.5 Conclusion

The main result of this chapter is that in a representative agent model with pollution generated by consumption satiation in consumption is always a sufficient condition for the downturn of the EKC and that a tendency to satiation – or in other words the normality of environmental quality – is even a necessary condition if we assume a standard functional form for the pollution function. The intuition is that if there is a tendency to satiation, growth in consumption does not increase utility by much and therefore additional income is devoted to abatement such that pollution falls. We have shown that *Stokey's* (1998) model is a special case of our model. With her specification of functional forms satiation is even all-decisive: With asymptotic satiation there is an EKC and without satiation pollution increases monotonically. This specification also shows that satiation can already cause pollution to fall at levels of consumption which are far below the satiation level.

This result has been derived employing our method developed in section 4.2, i.e. using only an ordinal, not a cardinal utility function. Moreover, we have allowed for a more general pollution function than in previous studies. This allowed us to state two im-

[18] *Andreoni* and *Levinson* (2001) restrict themselves to $\alpha = 1$ although their proof goes through for any $\alpha > 0$. Note that $P_C = \alpha - R_C < 0$ is possible in this model as $R_C > 0$.

portant conditions for the behaviour of pollution: The conditions whether or not consumption becomes increasingly less polluting and abatement increasingly more expensive with rising income while holding pollution constant. We have shown that at low income levels pollution rises with exogenously growing income because abatement expenditures are zero. At a higher income level, however, abatement expenditures become positive.

For further growing income the behaviour of pollution depends on satiation and the two conditions mentioned above. Pollution declines if the marginal rate of transformation (MRT) between pollution and consumption increases with income, for example if both conditions are violated. If zero pollution is technically feasible, pollution finally becomes zero (negative pollution is infeasible). If exactly one of the two conditions holds (for example if the cross derivative of the pollution function is zero as in previous models) and if there is a tendency to satiation, the pollution level also becomes small (or zero) at a sufficiently high income level. Thus we again find an EKC. Satiation, however, leads to an EKC independent of the two conditions and the MRT. Without satiation and if the two conditions hold, pollution can rise monotonically, it can become constant at some positive pollution level, or it can fall asymptotically approaching some level of pollution from above. The latter is the case if the marginal rate of substitution falls faster than the MRT (if we hold pollution constant), i.e. if there is a sufficiently strong tendency to satiation. Hence, only if the MRT increases with income, the tendency to satiation – or the normality of environmental quality – is not necessary to obtain an EKC. However, for standard functional forms the MRT decreases. All these results are independent of whether or not zero pollution is technically feasible.

It has been claimed by *McConnell* (1997) and other researchers that a high income elasticity of demand for environmental quality works in favour of falling pollution. We have shown, however, that a higher income elasticity can cause a more positive or a more negative slope of the PIR. Hence, there is no simple relationship between the income elasticity of demand for environmental quality and the PIR.

Building on the literature we presented a static model. This is unsatisfactory yet since the EKC is an inherently dynamic phenomenon in the course of economic growth and since the level of income needed to reach zero (or a low level of) pollution might be high. Thus it is not clear whether such a high income level is attained in a growth model. Furthermore, it is doubtful whether our results also apply in a dynamic framework. To analyse this we will study a dynamic version of the model in chapter 6.

Appendix: Counterexample to McConnell's claim

To determine the influence of the income elasticity of demand for environmental quality on the EKC, *McConnell* (1997) uses a counterfactual model in which the representative consumer can buy consumer goods at a price of one and environmental quality Q at a price of π, i.e. $Y = C + \pi Q$. Since pollution is the difference between the best environmental quality, Q_{max}, and the actual environmental quality, Q, the consumer maximizes[19]

$$\begin{aligned} \max_C U(C,P) &= \max_C U(C, Q_{max} - Q) \\ &= \max_C U(C, Q_{max} - (Y - C)/\pi). \end{aligned}$$

Calculating the first order condition we find $MRS = -U_C/U_P = 1/\pi$. This implies $MRS_C dC + MRS_P 1/\pi dC - MRS_P 1/\pi dY = 0$. Using $dC = dY - \pi dQ$ we derive

$$\frac{dQ}{dY} = \frac{MRS_C}{\pi MRS_C + MRS_P} > 0. \tag{5.15}$$

[19]Note that the consumer does not internalize the external effect he causes with his consumption.

as we have already stated in chapter 4 (see (4.3) and (4.4)). The income elasticity of demand for environmental quality is given by $\eta = Y/Q \, dQ/dY$.

McConnell's (1997, 391) claims that a higher η causes the slope of the PIR to become more negative (i.e. if the slope is positive, it becomes flatter and if it is negative, it becomes steeper). The slope of the PIR, dP/dY, is given by (5.9). We now present a counterexample to show that this claim is not correct. Taking the utility function (5.13) with $\epsilon = 0$ we derive $MRS = C^{-\sigma}/BP^{\gamma-1}$ and thus

$$MRS_C = -\tfrac{\sigma}{C} MRS, \qquad MRS_P = -\tfrac{\gamma-1}{P} MRS, \qquad \text{and}$$
$$\tfrac{dQ}{dY} = \tfrac{\sigma/C}{\pi\sigma/C+(\gamma-1)/P}.$$

Now we consider an optimal point, such as point X in figure 4.1 on page 104, where the indifference curve is tangent to the budget line, and we examine a change in preferences. Assume that γ falls and that B changes such that the $MRS = 1/\pi$ stays constant at the optimal point X.[20] It follows that MRS_C stays constant, but $|MRS_P|$ falls and dQ/dY rises. Since Q and Y remain unchanged, η increases too. Furthermore, the denominator of (5.9) decreases whereas the numerator stays constant. Thus if the slope of the PIR is positive, it becomes steeper. Hence, *McConnell's* (1997) claim that the slope becomes more negative when η rises is violated.

[20] In general a change in preferences also changes the optimal point and thus Q, C, and P.

6. A Dynamic Extension to McConnell's Model

6.1 Introduction

In chapter 5 income has been treated as an exogenous variable. This allowed us to work out some conditions for the EKC by simple comparative statics. A model of the EKC, however, should be dynamic as the EKC is an inherently dynamic phenomenon in the course of economic growth. In this chapter we endogenize growth by using the standard Ramsey approach. We analyse whether or not any results derived in the static model are changing in the more realistic dynamic framework. Surprisingly, we find almost the same results as in the static model. Therefore, the static model was also didactically useful since the intuition behind our results could be derived in a simpler framework. Furthermore, if pollution becomes zero at some high income level, we can examine whether or not this high income level is attained endogenously. Finally, while the dynamic model is more realistic, it also has disadvantages since it necessitates more rigorous assumptions.

We also study which results change with the elasticity of intertemporal substitution. To do so we apply the method developed in subsection 4.3.1, i.e. we analyse which results are robust to a monotonic transformation of the felicity function. We first derive as many results as possible using only an ordinal felicity function. Surprisingly, we obtain that almost all results are robust to monotonic transformations of the felicity function: The shape of the PIR – i.e. whether it is monotonically rising, an EKC, or N-shaped – is robust. The reason is that the pollution-consumption relationship is robust. If there is an EKC, the pollution level at the turning point is also robust. Only the income level at the turning point rises with the elasticity of intertemporal substitution.

This chapter is structured as follows. In section 6.2 we introduce the additional assumptions which are necessary to derive our results. In section 6.3 we show that the pollution-consumption relationships of the dynamic and of the static model are identical. After deriving the phase diagram in section 6.4, we analyse the PIR and how it changes with the elasticity of intertemporal substitution in section 6.5. Whether or not it is optimal to attain zero pollution is discussed in section 6.6. Section 6.7 concludes.

6.2 Additional assumptions

All the assumptions made in section 5.2 are also used in the model of this chapter except that we replace the static allocation of income $Y = C + A$ by the introduction of capital K: Taking the standard Ramsey approach we assume that production $\phi(K)$ has positive, but diminishing marginal returns, $\phi_K > 0$ and $\phi_{KK} < 0$. Income equals net output, i.e. $Y = \phi(K) - \delta K$ where δ is the depreciation rate. It can be used for consumption C, abatement expenditures A, or investment in new capital $\dot{K} = dK/dt$, i.e. $Y = C + A + \dot{K}$ or

$$\dot{K} = \phi(K) - \delta K - C - A.$$

The initial capital stock, K_0, is assumed to be small because this is the interesting case in which an EKC might emerge.

As mentioned earlier, we need two stronger assumptions to assure that the model is well-behaved, i.e. to ensure that the second order condition for a maximum is satisfied, that consumption never changes discontinuously, and that the steady state is a saddle point. First, we must assume that the Hessian $D^2H =$

$$\begin{bmatrix} H_{CC} & H_{AC} \\ H_{AC} & H_{AA} \end{bmatrix}$$ is negative definite where

$$\begin{aligned} H_{CC} &= U_{CC} + 2U_{CP}P_C + U_{PP}P_C^2 + U_P P_{CC} \\ H_{AC} &= H_{CA} = U_{CP}P_A + U_{PP}P_A P_C + U_P P_{AC} \\ H_{AA} &= U_{PP}P_A^2 + U_P P_{AA}. \end{aligned} \tag{6.1}$$

Therefore

$$H_{CC} < 0 \quad \text{and} \quad H_{AA}H_{CC} - H_{AC}^2 > 0 \tag{6.2}$$

both hold (see *Mas-Colell et al.* 1995, 936). These are mainly constraints on the permissible values of U_{CC}, U_{PP}, and U_{CP} – besides the constraints imposed by (4.1) and (4.2). In other words, these are constraints on the permissible monotonic transformations of the felicity function when studying whether or not the results are robust to changes in the elasticity of intertemporal substitution.

Second, if pollution becomes zero, we need two further assumptions: On the one hand, $U_{CC} \leq 0$ must hold, and on the other hand, the iso-pollution lines must be convex as in figure 5.1 on page 116. The curvature of the iso-pollution lines is determined by the derivative of the slope, $dA/dC = -P_C/P_A$, with respect to C. The iso-pollution lines are convex, if

$$\begin{aligned} \left.\frac{d^2A}{dC^2}\right|_{P=constant} &= \frac{-P_{CC}P_A - P_{AC}P_A dA/dC + P_{AA}P_C dA/dC + P_{AC}P_C}{P_A^2} \\ &= \frac{P_{CC} - 2P_{AC}P_C/P_A + P_{AA}P_C^2/P_A^2}{-P_A} \geq 0. \end{aligned} \tag{6.3}$$

In appendix A on page 257 it is shown that two standard functional forms fulfil (6.3) for all or for a vast array of parameter constellations. Note also that if P_{AC} were zero, (6.3) would be satisfied. Finally, U_{CC} and (6.3) are not allowed to be zero at the same time.

6.3 The pollution-consumption relationship

To find the first best allocation the social planer's problem is

$$\max_{C,A} \int_0^\infty e^{-\rho t} U(C, P(C,A))dt$$
$$\text{subject to} \quad \dot{K} = \phi(K) - \delta K - C - A \text{ and } A, P \geq 0,$$

where ρ is the time preference rate, t is time, and U is the felicity function. Note that pollution depends only on current values of consumption and abatement expenditures and is therefore a flow variable. The current value Lagrangian L is defined by

$$L = U(C, P(C,A)) + \lambda(\phi(K) - \delta K - C - A) + \gamma A + \mu P(C,A)$$

where λ is the costate variable and γ and μ are the multipliers of the nonnegativity constraints. The Hamiltonian H is the same without $\gamma A + \mu P$ (*Feichtinger* and *Hartl* 1986, 161). So the first order conditions are

$$L_C = U_C + (U_P + \mu)P_C - \lambda = 0 \tag{6.4}$$
$$L_A = (U_P + \mu)P_A - \lambda + \gamma = 0 \tag{6.5}$$
$$-L_K = -\lambda(\phi_K - \delta) = \dot{\lambda} - \rho\lambda. \tag{6.6}$$

The transversality condition is

$$\lim_{t\to\infty} e^{-\rho t}\lambda(t)[K(t) - \check{K}(t)] \geq 0 \tag{6.7}$$

where $\check{K}(t)$ is capital on the optimal path and $K(t)$ is capital on any other feasible path. (6.4) and (6.5) are identical to (5.3) and (5.4) from which we again derive

$$MRS = -\frac{U_C}{U_P} = (P_C - P_A)\frac{U_P + \mu}{U_P} - \frac{\gamma}{U_P} = MRT\frac{U_P + \mu}{U_P} - \frac{\gamma}{U_P} \tag{6.8}$$

which is identical to (5.6). So as in the static model the indifference curve must be tangent to the consumption possibilities curve (CPC) at an interior solution and we conclude:

Proposition 6.1 *The pollution-consumption relationship of the dynamic model is exactly identical to that of the static version of the model.*

The reason is that only (5.6) was used to derive the pollution-consumption relationship in the static model. A formal proof is given in the appendix to this chapter. Whenever a certain level of consumption is chosen in the dynamic model, abatement expenditures and pollution take the values implied by the optimal point in figure 5.2 on page 118 at that level of consumption. Hence, abatement expenditures are zero at low levels of consumption, but become positive at point B in figure 5.2. Let us denote the consumption level at point B by $\tilde{C}$. Thus for $C \leq \tilde{C}$ we have $A = 0$ and pollution rises with consumption. At an interior solution abatement expenditures rise with consumption. Furthermore, pollution falls with rising consumption if $-MRS_C > W - V$ (see (5.8)) or in other words if on a horizontal ray in figure 5.2 the slope of the indifference curve, i.e. the MRS, falls faster than the slope of the CPC, i.e. the MRT. Hence, for standard functional forms for the pollution function it is necessary for an EKC to exist that environmental quality is a normal good or in other words that there is a tendency to satiation (see corollary 5.2). If the indifference curves become flatter than the $CPCs$ on the C-axis, pollution becomes zero (see (5.10)). This happens at point G in figure 5.2. The consumption level at point G is denoted by $C_{P=0}$. Thus for $C \geq C_{P=0}$ we have $P = 0$. As shown in proposition 5.3 pollution becomes zero if zero pollution is technically feasible and if there is satiation or if there is a tendency to satiation and either condition (5.1) or (5.2) is violated, i.e. if – holding pollution constant – either consumption becomes increasingly more polluting or abatement becomes increasingly less expensive. If zero pollution requires infinite abatement expenditures, we find that under the same conditions pollution approaches zero as consumption goes to infinity or to the satiation level (see corollary 5.4).

Andreoni and *Levinson's* (2001) result holds also true in the dynamic model because their proof also rests on figure 5.2: If there are increasing returns to scale in abatement, pollution becomes zero at a sufficiently large consumption level.

6.4 Derivation of the phase diagram

So far we have only derived the pollution-consumption relationship, but our aim is to derive the pollution-income relationship (PIR). In the static model consumption increases with income so that these two relationships have the same shape. To analyse the PIR in the dynamic model we must first derive the phase diagram. The phase diagram is three-dimensional with consumption, abatement expenditures, and capital on the three axes. Using the results from section 6.3, however, we can make it two-dimensional: We always choose the optimal level of abatement for a given level of consumption according to the pollution-consumption relationship in figure 5.2. By means of this procedure we are able to reduce the maximization problem with two control variables (C and A) into a problem with only one control variable (C).

As shown in section 6.3 abatement expenditures are zero for $C \leq \tilde{C}$, i.e. below the $\tilde{C}$-line in figures 6.1 and 6.2, and pollution is zero for $C \geq C_{P=0}$, i.e. above the $C_{P=0}$-line in figure 6.2. There is no $C_{P=0}$-line if zero pollution is technically not feasible or if the pollution-consumption relationship is monotonically increasing. To complete the phase diagram we need to derive the $\dot{K} = 0$- and the $\dot{C} = 0$-line. The corresponding lines for the case of the standard Ramsey model without pollution (i.e. with $P = P_C = P_A = 0$) are denoted by $\dot{\bar{K}} = 0$- and $\dot{\bar{C}} = 0$-line. All these lines are shown in figures 6.1 and 6.2.

Lemma 6.1 *1. The $\dot{K} = 0$- and the $\dot{\bar{K}} = 0$-line are both rising (falling) for $K < K^*$ ($K > K^*$) where $\phi_K(K^*) = \delta$. For $C \leq \tilde{C}$ they are identical. When the $\dot{K} = 0$-line crosses the*

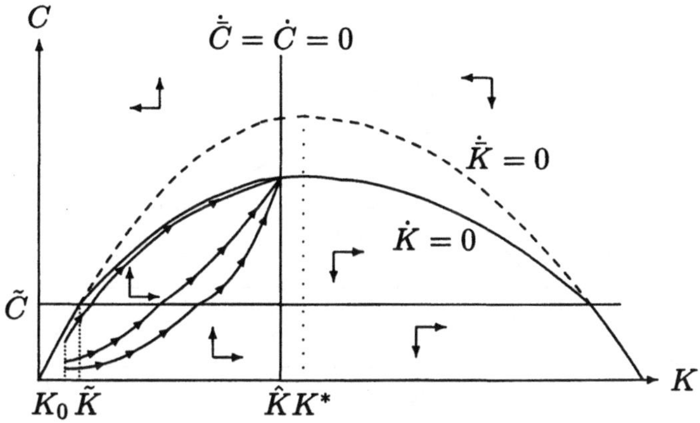

Figure 6.1: Phase diagram with three possible paths.

$\tilde{C}$-line ($C_{P=0}$-line) from below, it becomes suddenly flatter (steeper). Below (above) the $\dot{K} = 0$-line capital rises (falls).

2. *The $\dot{C} = 0$- and the $\dot{\tilde{C}} = 0$-line are identically vertical at $\hat{K} < K^*$ where $\phi_K(\hat{K}) = \rho + \delta$. Consumption rises (falls) to the left (right) of $\hat{K}$.*

The proof is relegated to the appendix. The steady state lies at the intersection of the $\dot{K} = 0$- and the $\dot{C} = 0$-line. Without pollution we would find the same capital stock in the steady state: Since capital is not polluting, there is no reason to stop capital growth at a smaller capital stock in our model as compared to the standard Ramsey model. However, the steady state lies at a smaller consumption level since part of output is used for abatement: The vertical distance between the $\dot{\tilde{K}} = 0$- and the $\dot{K} = 0$-line are the (optimally chosen) abatement expenditures because these lines are defined by $\dot{\tilde{K}} = \phi(K) - \delta K - C = 0$ and $\dot{K} = \phi(K) - \delta K - C - A = 0$, respectively. Therefore the $\dot{K} = 0$-line is kinked at $\tilde{C}$ and $C_{P=0}$: Since the abatement expenditures are positive above the $\tilde{C}$-line, the $\dot{K} = 0$-line lies below the $\dot{\tilde{K}} = 0$-line. Above $C_{P=0}$ negative pollution would be optimal. As this is not feasible, we need less

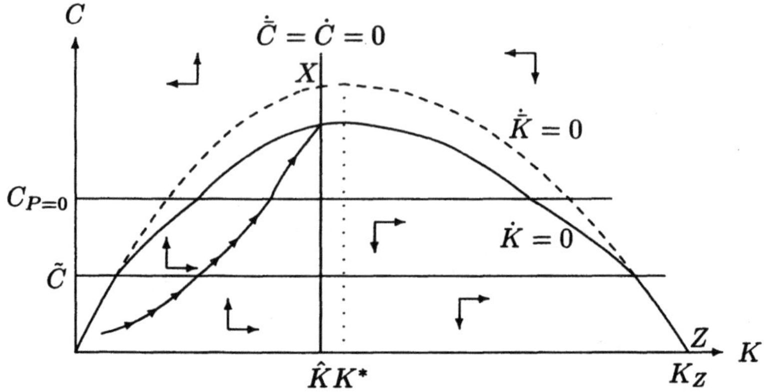

Figure 6.2: Phase diagram with zero pollution steady state.

abatement to have zero pollution instead of negative. So the vertical distance between the $\dot{\hat{K}} = 0$- and the $\dot{K} = 0$-line is smaller than it would be if negative pollution was feasible.

Before we proceed with the analysis of the phase diagram we need a few further results.

Lemma 6.2

1. *The steady state is a saddle point.*
2. *Discontinuous changes in consumption are never optimal.*
3. *The path into the saddle point is the only path that satisfies both the first order conditions (6.4) – (6.6) and the transversality condition (6.7).*
4. *The second order conditions for a maximum are satisfied.*

The proof can be found in the appendix. That the steady state is a saddle point can also be seen in figures 6.1 and 6.2. It is also important to prove that it is never optimal to change consumption, i.e. the control variable, discontinuously at some point in time. Finally, it is good to know that the saddle path is the only path that satisfies all necessary and sufficient conditions for a maximum.

Figure 6.2 shows the optimal path of the economy if pollution becomes zero. At $\tilde{C}$ the path is kinked since part of output is now used for abatement. It is also kinked at $C_{P=0}$ because a greater part of output would be used for abatement above $C_{P=0}$ if negative pollution was feasible.

6.5 The pollution-income relationship

We are now in a position to derive the PIR. Income, $Y = \phi(K) - \delta K$, (or GDP $= \phi(K)$ which is used in empirical studies) increases with capital along the optimal path (income would reach its maximum at the maximum of the $\dot{K} = 0$-line which is beyond the steady state). Therefore for each given level of income (or GDP) the corresponding capital stock follows from $Y = \phi(K) - \delta K$ (or GDP $= \phi(K)$). Next, for each given capital stock the optimal path in figure 6.1 or 6.2 determines the level of consumption. For each given level of consumption, in turn, the pollution-consumption relationship in figure 5.2 (see page 118) determines the optimal level of pollution. With this procedure the PIR can be derived.

However, we now show that the PIR is not robust to changes in the elasticity of intertemporal substitution. As shown in section 4.3 a change in the elasticity of intertemporal substitution is a monotonic transformation of the felicity function.

Lemma 6.3 *We can decrease the elasticity of intertemporal substitution such that the optimal path in the phase diagram shifts upwards whereas $\tilde{C}$, $C_{P=0}$, the $\dot{K} = 0$-, and the $\dot{C} = 0$-line are all unaffected.*

The proof is given in the appendix. Three paths for three monotonic transformations of the same ordinal felicity function are drawn in figure 6.1. Consumption and capital always grow along the optimal path. However, a monotonic transformation changes the curvature of the optimal path and thus the level of consumption for a given capital stock on the optimal path: The elasticity

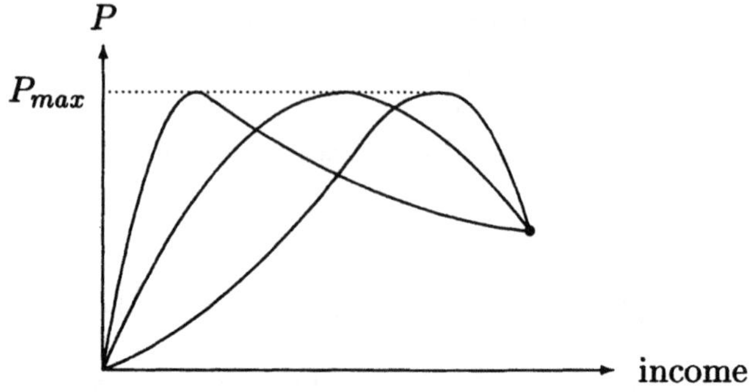

Figure 6.3: The PIR for different monotonic transformations of the felicity function.

of intertemporal substitution determines whether it is optimal to keep consumption at low levels to reach the steady state early on due to fast capital growth or whether it is better to have high consumption at low capital stocks and to let capital grow slowlier. *Barro* and *Sala-i-Martin* (1995, 76) derive a similar result for the standard Ramsey model without pollution.

For the next proposition we need a bit of notation: We define $\tilde{K}$ as the capital stock at the intersection of the $\tilde{C}$- and the $\dot{K} = 0$-line as shown in figure 6.1. Moreover, we denote by PIR_U and $\text{PIR}_{\tilde{U}}$ the optimal PIR for the felicity function U and for any monotonic transformation $\tilde{U}$ of the felicity function U, respectively.

Proposition 6.2 *1. Suppose that $K_0 < \tilde{K}$. Then if PIR_U is monotonically rising (or an EKC or N-shaped), $PIR_{\tilde{U}}$ is also monotonically rising (or an EKC or N-shaped). Furthermore, PIR_U and $PIR_{\tilde{U}}$ both have the same steady state level of income and pollution, and if PIR_U and $PIR_{\tilde{U}}$ are EKCs, the pollution levels at the turning points of the EKCs are equal (see figure 6.3).*

2. *If the PIR is an EKC, the income level at the turning point of the EKC rises with the elasticity of intertemporal substitution (see figure 6.3).*

Proof. 1. Starting at $K_0 < \tilde{K}$ capital (and thus income $Y = \phi(K) - \delta K$) and consumption are growing on every path which is optimal for some felicity function (see figure 6.1). Since $K_0 < \tilde{K}$, abatement expenditures are first set equal to zero on PIR_U and $\text{PIR}_{\tilde{U}}$ such that both are first rising. Both PIR_U and $\text{PIR}_{\tilde{U}}$ follow the pollution-consumption relationship shown in figure 5.2 (see page 118) which is robust to a monotonic transformation of the felicity function (see (6.8)). Hence, the shape of the PIR (i.e. whether it is monotonically rising, an EKC, or N-shaped) and the pollution level at the turning point of the EKC – if there is one – are determined by the pollution-consumption relationship only and are thus robust to changes in the elasticity of intertemporal substitution. Furthermore, since the $\dot{K} = 0$- and the $\dot{C} = 0$-line are robust (see lemma 6.3), the steady state is also robust.

2. Consider the three paths in figure 6.1 whose PIRs are drawn in figure 6.3 (assuming that the PIR is an EKC). According to lemma 6.3 a higher path is associated with a smaller elasticity of intertemporal substitution. On the three paths in figure 6.1 the level of consumption chosen for a given level of capital and thus of income is different (for $K < \hat{K}$). Therefore the consumption level associated with the highest pollution level (see figure 5.2) is reached at a different capital and thus income level on the three paths: The income level at the turning points of the EKC rises with the elasticity of intertemporal substitution. □

Hence, the shape of the PIR – i.e. whether it is monotonically rising, an EKC, or N-shaped – and the pollution level at the turning point of the EKC are robust to changes in the elasticity of intertemporal substitution. In contrast, the income level at the turning point of the EKC is not robust. As mentioned in subsection 2.3.2, the pollution level at the turning point is more important than the income level since it determines whether or not (irreversible) damages occur (*Agras* and *Chapman* 1999, 276, *Borghesi* 1999, and *Perrings* and *Ansuategi* 2000). Hence, all crucial results are robust to monotonic transformations of the felicity function.

6.6 Is zero pollution attained?

The last question we want to answer in this chapter is under what conditions zero pollution is attained. To do so we assume that zero pollution is technically feasible and we analyse whether or not it is optimal to reach zero pollution. Infinite growth is not possible as the economy only grows until net marginal output equals the time preference rate, i.e. until $\phi_K - \delta = \rho$. So in this model it is not the environment that stops capital and thus income growth. Zero pollution is reached if the following conditions all hold simultaneously:

- If the steady state income level is high. The steady state income level is given by $Y = \phi(\hat{K}) - \delta\hat{K}$ where $\phi_K(\hat{K}) = \rho + \delta$ or by the consumption level at point X in figure 6.2. It is high if capital is highly productive ($\phi(K)$ high) and has a high marginal productivity (ϕ_K always high implying $|\phi_{KK}|$ small) and if the rates of depreciation and of time preference are small (δ and ρ small). Of course, a high income level is not sufficient for zero pollution. Additionally, zero pollution must be reached at a low income level. This is the case

- if pollution becomes zero at a small consumption level. According to (5.10) this is the case if on the C-axis in figure 5.2 on page 118
  - the indifference curves are flat and become quickly flatter with rising consumption. Then the MRS is small and $|MRS_C|$ is high. A tendency to satiation (U_C small and falling with rising C) or 'green' preferences ($|U_P|$

high and rising with C) cause the indifference curves to be flat and become flatter.

- the consumption possibilities curves ($CPCs$) are steep and become quickly steeper with rising consumption, i.e. if consumption is pollution-intensive and becomes increasingly more so (P_C high and (5.1) violated) and if abatement is cheap and becomes increasingly cheaper ($|P_A|$ high and (5.2) violated). If (5.1) and (5.2) are satisfied, the CPC becomes flatter with rising consumption. Then the MRS must fall faster than the CPC on the C-axis in figure 5.2.

If these two conditions are violated, (5.8) does neither hold and pollution rises monotonically such that zero pollution is never optimal. For the sake of the argument we assume that pollution becomes zero at some consumption level. Then abatement expenditures might be extremely high such that zero pollution requires a higher income level than we can reach. Hence, zero pollution is only attained,

- if, additionally, abatement expenditures are small on the C-axis in figure 5.2, or in other words
  - if consumption is only slightly polluting ($P(C, 0)$ small) and, of course,
  - if abatement is cheap ($|P_A|$ always high, implying P_{AA} small), i.e. if we need only little money to abate all pollution caused by consumption or, in other words, if abatement expenditures must only increase slightly to move from point Z to point G in figure 5.2 on page 118.

Therefore a small P_C means that the level of consumption where pollution becomes zero is higher, but it also means that less abatement expenditures are necessary to reach zero pollution. Thus the effect of P_C on the level of income necessary for zero pollution is ambiguous.

Hence, if zero pollution is optimal at some high consumption level, capital must be sufficiently productive and abatement sufficiently cheap to actually reach zero pollution. If they are not, it could be asked whether or not abatement expenditures are positive in the steady state. This depends on similar reasons: Abatement expenditures are positive in the steady state if the steady state income level is high and if the indifference curves become quickly flatter than the CPCs on the $P(C, 0)$-line in figure 5.2, i.e. if the MRS is small and P_C and $|P_A|$ are high. If zero pollution is technically not feasible, the steady state pollution level is small under the same conditions as derived above.

6.7 Conclusion

The results in the dynamic and in the static model are almost identical. However, in the dynamic model income stops growing. The steady state income level is sufficiently high to have zero pollution – or at least to have positive abatement expenditures and a low pollution level – if the production technology is highly productive. We have been preoccupied by analysing under what conditions pollution becomes zero. As shown in chapter 5 (see (5.8)), however, it is also possible that pollution rises monotonically.

An important result is that the pollution-consumption relationship in the dynamic model is exactly the same as in the static version of the model. So the explanations of the EKC derived in the static framework of chapter 5 are also valid in the dynamic extension. It also follows that the shape of the PIR (i.e. whether it is monotonically rising, an EKC, or N-shaped) does not change with the elasticity of intertemporal substitution. Similarly, if there is an EKC, the pollution level at the turning point of the EKC is determined by the pollution-consumption relationship and is thus robust to a monotonic transformation of the felicity function. Nevertheless, the results for the PIR are more complicated because the income level at the turning point of the EKC rises with the elasticity of intertemporal substitution. However, the pollution level at the turning point is more important than the income level because it determines what damages occur. Hence, all crucial results are robust. Following the method developed in section 4.3 we have first derived as many results as possible (the pollution-consumption relationship) using only ordinal assumptions about the felicity function. Only then we went on to derive those results which additionally require cardinal assumptions.

Appendix: Proofs

Proof of proposition 6.1

It follows directly from the proof of proposition 5.1 that abatement expenditures are zero at low levels of consumption.

To prove that abatement expenditures are increasing with consumption at an interior solution we need an additional intermediate result. By taking the derivatives of (6.4) and (6.5) with respect to C and A at an interior solution ($\gamma = \mu = 0$) we see that the second derivatives of the Hamiltonian H are given in (6.1). Note that at an interior solution[1]

$$\begin{aligned} (H_{AA} - H_{AC})/U_P &= MRS_P P_A + P_{AA} - P_{AC} > 0 \\ (H_{CC} - H_{AC})/U_P &= -MRS_C - MRS_P P_C + P_{CC} - P_{AC} > 0. \end{aligned} \tag{6.9}$$

Differentiating (6.4) and (6.5) with respect to time we find

[1] On the one hand, inserting (4.2) and $U_C/U_P = P_A - P_C$ which is (6.8) at an interior solution we find

$$\begin{aligned} MRS_P P_A + P_{AA} - P_{AC} &= (-U_{CP}U_P P_A + U_{PP}U_C P_A)/U_P^2 + P_{AA} - P_{AC} \\ &= (-U_{CP}P_A + U_{PP}P_A^2 - U_{PP}P_A P_C + U_P P_{AA} - U_P P_{AC})/U_P \\ &= (H_{AA} - H_{AC})/U_P. \end{aligned}$$

On the other hand, we derive from (4.1), (4.2), and (6.8) that $-MRS_C -$

$$H_{CC}\dot{C} + H_{AC}\dot{A} + \mu P_{CC}\dot{C} + \mu P_{AC}\dot{A} + \dot{\mu} P_C - \dot{\lambda} = 0$$
$$H_{AA}\dot{A} + H_{AC}\dot{C} + \mu P_{AA}\dot{A} + \mu P_{AC}\dot{C} + \dot{\mu} P_A - \dot{\lambda} + \dot{\gamma} = 0. \quad (6.10)$$

Assuming an interior solution ($\dot{\gamma} = \mu = \dot{\mu} = 0$) and inserting (6.9) we conclude

$$\begin{aligned} \dot{A} &= \dot{C}\frac{H_{CC} - H_{AC}}{H_{AA} - H_{AC}} \quad (6.11) \\ &= \dot{C}\frac{-MRS_C - MRS_P P_C + P_{CC} - P_{AC}}{MRS_P P_A + P_{AA} - P_{AC}} > 0 \text{ for } \dot{C} > 0. \end{aligned}$$

Thus dA/dC is the fraction in (6.11). It is identical with what is found in the static model (see proof of proposition 5.2 where $dA/dC = -b_1/b_2$). If pollution is zero, dA/dC is also identical since we find from $dP = P_C dC + P_A dA = 0$ that

$$dA/dC = -P_C/P_A > 0. \quad (6.12)$$

At an interior solution we insert (6.11) into $\dot{P} = P_C\dot{C} + P_A\dot{A}$ to find

$$\begin{aligned} \dot{P} &= \dot{C}\,\frac{P_C(MRS_P P_A + P_{AA} - P_{AC}) + P_A(-MRS_C - MRS_P P_C + P_{CC} - P_{AC})}{MRS_P P_A + P_{AA} - P_{AC}} \\ &= \dot{C}\,\frac{P_{CC}P_A - P_{AC}P_C - P_{AC}P_A + P_{AA}P_C - MRS_C P_A}{MRS_P P_A + P_{AA} - P_{AC}} \end{aligned}$$

where the denominator is positive. For $\dot{C} > 0$ the sign of $\dot{P}$ depends on the numerator which is exactly identical with the numerator of (5.9). So pollution falls with rising consumption, if (5.8) holds.

$MRS_P P_C + P_{CC} - P_{AC}$

$$\begin{aligned} &= (U_{CC}U_P - U_{PC}U_C + U_{CP}U_P P_C - U_{PP}U_C P_C)/U_P^2 + P_{CC} - P_{AC} \\ &= (U_{CC} - U_{PC}P_A + 2U_{PC}P_C - U_{PP}P_A P_C + U_{PP}P_C^2 + U_P P_{CC} \\ &\quad - U_P P_{AC})/U_P = (H_{CC} - H_{AC})/U_P. \end{aligned}$$

Since (6.8) is identical with (5.6), the proofs of proposition 5.3 and corollary 5.4 go through. □

Proof of lemma 6.1

1. From $\dot{K} = \phi(K) - \delta K - C - A(C) = 0$ we derive

$$\left.\frac{dC}{dK}\right|_{\dot{K}=0} = -\frac{\partial\dot{K}/\partial K}{\partial\dot{K}/\partial C} = \frac{\phi_K - \delta}{1 + dA/dC} \gtrless 0 \text{ if } K \lessgtr K^*. \quad (6.13)$$

For $C < \tilde{C}$ and for the $\dot{\tilde{K}} = 0$-line $A = dA/dC = 0$, for an interior solution ($\tilde{C} < C < C_{P=0}$) dA/dC is given in (6.11), and for $C > C_{P=0}$ dA/dC is given in (6.12). Thus for $C \leq \tilde{C}$ the $\dot{K} = 0$- and the $\dot{\tilde{K}} = 0$-line are identical. Coming from below the $\dot{K} = 0$-line suddenly becomes flatter at $\tilde{C}$ since dA/dC suddenly becomes positive. At $C_{P=0}$ the $\dot{K} = 0$-line suddenly becomes steeper because dA/dC suddenly becomes smaller (it is easy to show that (6.12) is smaller than the fraction in (6.11) because (5.8) holds at $C_{P=0}$ since pollution must be falling to reach zero pollution). Below (above) the $\dot{K} = 0$-line C – and thus $A(C)$ – is smaller (greater) than on the line. So $\dot{K} = \phi(K) - \delta K - C - A(C)$ is positive (negative) as indicated by the arrows in figures 6.1 and 6.2.

2. For $C \leq \tilde{C}$ or for the $\dot{\tilde{C}} = 0$-line $A = \dot{A} = \mu = \dot{\mu} = 0$ holds. Combining the first line of (6.10) with (6.6) and (6.5) or (6.4) it follows that

$$\dot{C} = \frac{U_P P_A + \gamma}{H_{CC}}(\rho+\delta-\phi_K) \quad \text{or} \quad \dot{C} = \frac{U_C + U_P P_C}{H_{CC}}(\rho+\delta-\phi_K). \quad (6.14)$$

For the $\dot{\tilde{C}} = 0$-line the latter formula with $P_C = 0$ and $H_{CC} = U_{CC}$ (see (6.1)) is more appropriate. The first formula shows that the fraction is negative (see (6.2)). So the $\dot{C} = 0$-line for $C \leq \tilde{C}$ and the $\dot{\tilde{C}} = 0$-line are vertical at $\hat{K}$ where $\phi_K(\hat{K}) = \rho + \delta$.

For $\tilde{C} < C < C_{P=0}$ we insert $\mu = \dot{\mu} = 0$ and (6.11) into the first line of (6.10) to find $\dot{C}\{H_{CC}(H_{AA} - H_{AC}) + H_{AC}(H_{CC} -$

$H_{AC})\}/\{H_{AA} - H_{AC}\} = \dot{\lambda}$. Together with (6.6) and (6.5) we obtain

$$\dot{C} = \frac{U_P P_A (H_{AA} - H_{AC})}{H_{AA} H_{CC} - H_{AC}^2} (\rho + \delta - \phi_K) \tag{6.15}$$

where the fraction is negative due to (6.2) and (6.9). Thus for $\tilde{C} < C < C_{P=0}$ the $\dot{C} = 0$-line is also vertical at $\hat{K}$.

For $C \geq C_{P=0}$ we must first derive $\dot{\mu}$. It follows from (6.10) and (6.12) that

$$\begin{aligned} &(H_{CC} + \mu P_{CC})\dot{C} - (H_{AC} + \mu P_{AC})\frac{P_C}{P_A}\dot{C} + \dot{\mu} P_C \\ &= \; -(H_{AA} + \mu P_{AA})\frac{P_C}{P_A}\dot{C} + (H_{AC} + \mu P_{AC})\dot{C} + \dot{\mu} P_A \end{aligned}$$

and thus inserting (6.1) yields

$$\begin{aligned} \dot{\mu}(P_C - P_A) = \dot{C}[&- U_{CC} - 2U_{CP}P_C - U_{PP}P_C^2 - (U_P + \mu)P_{CC} \\ &+U_{CP}P_C + U_{PP}P_C^2 + (U_P + \mu)P_{AC}P_C/P_A - U_{PP}P_A P_C \\ &-(U_P + \mu)P_{AA}P_C/P_A + U_{CP}P_A + U_{PP}P_A P_C + (U_P + \mu)P_{AC}]. \end{aligned}$$

Cancelling terms we find

$$\dot{\mu} = \dot{C} \, \frac{-(U_P+\mu)(P_{CC}P_A - P_{AC}P_C - P_{AC}P_A + P_{AA}P_C) - U_{CC}P_A + U_{CP}P_A(P_A - P_C)}{P_A(P_C - P_A)}.$$

Next, from the first line of (6.10), (6.12), and (6.1) it follows that

$$\begin{aligned} \dot{\lambda} &= (H_{CC} + \mu P_{CC})\dot{C} - (H_{AC} + \mu P_{AC})P_C/P_A\dot{C} + \dot{\mu} P_C \\ &= \dot{C}[U_{CC} + U_{CP}P_C + (U_P + \mu)(P_{CC} - P_{AC}P_C/P_A)] + \dot{\mu} P_C. \end{aligned}$$

Inserting $\dot{\mu}$ we obtain

$$\begin{aligned} \dot{\lambda} = \dot{C}\tfrac{1}{P_A(P_C - P_A)} \Big[&U_{CC}P_A P_C - U_{CC}P_A^2 + U_{PC}P_C P_A(P_C - P_A) \\ &+(U_P + \mu)P_A P_C P_{CC} - (U_P + \mu)P_A^2 P_{CC} - (U_P + \mu)P_{AC}P_C^2 \\ &+(U_P + \mu)P_{AC}P_A P_C - (U_P + \mu)(P_{CC}P_A P_C - P_{AC}P_C^2 \\ &-P_{AC}P_A P_C + P_{AA}P_C^2) - U_{CC}P_A P_C + U_{CP}P_A P_C(P_A - P_C)\Big] \\ = \dot{C}\tfrac{P_A}{P_C - P_A}\Big[&-U_{CC} \\ &+(U_P + \mu)(-P_{CC} + 2P_{AC}P_C/P_A - P_{AA}P_C^2/P_A^2)\Big]. \end{aligned} \tag{6.16}$$

Inserting $\dot{\lambda}$ from (6.6) and (6.5) and $U_P + \mu = -U_C/(P_C - P_A)$ from (6.8) we derive

$$\dot{C} = \frac{P_C - P_A}{(P_C - P_A)U_{CC}/U_C - P_{CC} + 2P_{AC}P_C/P_A - P_{AA}P_C^2/P_A^2}(\rho + \delta - \phi_K). \tag{6.17}$$

The fraction is negative due to our assumption that U_{CC} and (6.3) are never simultaneously zero. Thus for $C \geq C_{P=0}$ the $\dot{C} = 0$-line is also vertical at $\hat{K}$.

To the left (right) of the $\dot{C} = 0$-line $\rho + \delta - \phi_K < 0$ (> 0) holds and thus consumption grows (shrinks) as shown by the arrows in figure 6.1 and 6.2 since the fractions in (6.14), (6.15), and (6.17) are all negative. □

Proof of lemma 6.2

1. The type of the steady state (SS) can be determined with the Jacobian matrix J. If the steady state lies between $\tilde{C}$ and $C_{P=0}$, we denote the fraction in (6.15) by $S < 0$ and we use $dA/dC > 0$ from (6.11). From $\dot{K} = \phi(K) - \delta K - C - A(C)$ and (6.15) we obtain

$$\left.\frac{\partial \dot{K}}{\partial K}\right|_{SS} = \phi_K - \delta = \rho > 0 \qquad \left.\frac{\partial \dot{K}}{\partial C}\right|_{SS} = -1 - dA/dC < 0$$
$$\left.\frac{\partial \dot{C}}{\partial K}\right|_{SS} = -S\phi_{KK} < 0 \qquad \left.\frac{\partial \dot{C}}{\partial C}\right|_{SS} = \partial S/\partial C(\rho + \delta - \phi_K) = 0$$

where all derivatives are evaluated at the steady state where $\phi_K = \delta + \rho$ as $\dot{C} = 0$. Thus

$$J_{SS} = \begin{bmatrix} \rho > 0 & -1 - dA/dC < 0 \\ -S\phi_{KK} < 0 & 0 \end{bmatrix}$$

and $\det J_{SS} < 0$. Hence, the steady state is a saddle point as shown in figure 6.1 (*Chiang* 1984, 643). If the steady state lies above $C_{P=0}$, we only have to redefine S as the fraction in (6.17) and use (6.12) instead of (6.11). The calculations are not changed by this, so the steady state is also a saddle point as shown in figure

6.2. Finally, if the steady state lies below $\tilde{C}$, it is also a saddle point since we again only have to redefine S as the fraction in (6.14) and use $dA/dC = 0$.

2. On an optimal path λ must be a continuous function of time (*Feichtinger* and *Hartl* 1986, 40). Thus sudden jumps of the control variable C, say from C_1 to C_2, can only be optimal, if λ takes the same value at C_1 and C_2. This is never the case because λ decreases steadily with rising consumption: According to (6.4) $\lambda = U_C + (U_P + \mu)P_C$. Thus for $C < \tilde{C}$ we find $d\lambda/dC = H_{CC}$ which is negative due to (6.2). For $\tilde{C} < C < C_{P=0}$ we obtain $d\lambda/dC = H_{CC}+H_{AC}dA/dC = (H_{AA}H_{CC}-H_{AC}^2)/(H_{AA}-H_{AC}) < 0$ where we have inserted (6.11) and where the sign follows from (6.9) and (6.2). Finally, for $C > C_{P=0}$ it follows immediately from (6.16) that $d\lambda/dC < 0$.

3. It is clear from its derivation that the saddle path satisfies the first order conditions (6.4) – (6.6). To show that the transversality condition (6.7) also holds, we follow *Barro* and *Sala-i-Martin* (1995, 65) and use (6.6) to write $\lambda(t)$ as

$$\lambda(t) = \lambda(0) \exp\left\{ \int_0^t \rho + \delta - \phi_K(\check{K}(t))dt \right\}$$

where $\check{K}$ is capital on the optimal path. Hence, the transversality condition (6.7) becomes

$$\lim_{t\to\infty} \lambda(0)[K(t) - \check{K}(t)] \exp\left\{ \int_0^t \delta - \phi_K(\check{K}(t))dt \right\} \geq 0.$$

The saddle path lies always to the left of $\hat{K}$ where $\phi_K(\hat{K}) = \rho + \delta$ (see figure 6.1 or 6.2). So $\delta - \phi_K \leq -\rho$ holds along the whole saddle path. Hence, the transversality condition is satisfied, because $K - \check{K}$ is bounded such that the limit is zero.

Next, we show that every other path violates either the first order conditions or the transversality condition. A path can only be optimal when it satisfies the first order conditions (6.4) – (6.6). Hence, in figure 6.1 or 6.2 we must follow the arrows which have been derived from (6.4) – (6.6). On the one hand, if – starting at K_0 – we choose a higher consumption level than on the saddle path, the economy sooner or later reaches the C-axis at some positive level of consumption (cf. figure 6.2). Since capital and consumption must both be nonnegative, the economy must then jump into the origin ($C = K = 0$). Because we have shown – based on the first order conditions – that jumping is never optimal, such a path violates the first order conditions.[2] On the other hand, if we start with a smaller consumption level than on the saddle path, the economy ends up in point Z in figure 6.2 where Z is defined as the point $(K_Z, 0)$ where $\phi(K_Z) = \delta K_Z$. For all these paths the transversality condition is violated, first, because $\lambda(0) > 0$ which follows from (6.5) since $P(0) > 0$ and thus $\mu(0) = 0$, second, because $\lim_{t\to\infty}[K(t) - \check{K}(t)] = \hat{K} - K_Z < 0$ if we compare a path going into point Z with the saddle path, and, third, because a path going into point Z lies most of the time at capital stocks $K > K^*$ where $\phi_K(K^*) = \delta$ (see figure 6.2) such that $\delta > \phi_K$ most of the time and thus $\lim_{t\to\infty} \exp\left\{\int_0^t \delta - \phi_K(\check{K}(t))dt\right\} = \infty$.

4. To prove that the sufficient conditions for a maximum are satisfied, we must show that H^0 is concave in K where H^0 is the Hamiltonian with the optimal values of C and A inserted for a given K (*Feichtinger* and *Hartl* 1986, 43). Note first that $\lambda \geq 0$ always holds ((6.8) shows that $U_P + \mu \leq 0$ such that $\lambda \geq 0$ follows from (6.5)). For zero abatement expenditures we find $H^0 = U\{C(K), P(C(K), 0)\} + \lambda(\phi(K) - \delta K - C(K))$. This yields $H_K^0 = \lambda(\phi_K - \delta) + H_C dC/dK$ where $H_C = 0$ is (6.4) with $\mu = 0$. Since $\phi_{KK} < 0$ and $H_{CC} < 0$ (see (6.2)), we conclude that H^0 is concave as

$$H_{KK}^0 = \lambda\phi_{KK} + H_{CC}(dC/dK)^2 < 0.$$

[2] Alternatively, we could argue that when the economy reaches the C-axis, consumption would have to grow according to (6.14), (6.15), or (6.17) which have been derived from (6.4) – (6.6). However, consumption must fall, since capital and consumption must be nonnegative. Therefore, the necessary conditions for an optimal path (6.4) – (6.6) are not satisfied on a path reaching the C-axis.

At an interior solution we obtain $H^0 = U\{C(K), P[C(K), A(C(K))]\} + \lambda[\phi(K) - \delta K - C(K) - A(C(K))]$. Thus $H^0_K = \lambda(\phi_K - \delta) + H_C dC/dK + H_A dA/dC\ dC/dK$ where $H_C = 0 = H_A$ are given in (6.4) and (6.5) with $\mu = \gamma = 0$. Therefore H^0_{KK} equals

$$\lambda\phi_{KK} + (dC/dK)^2\left(H_{CC} + 2H_{AC}dA/dC + H_{AA}(dA/dC)^2\right).$$

Because D^2H is negative definite (see (6.1)), $z'[D^2H]z < 0$ holds for all vectors $z \neq 0$ (*Mas-Colell et al.* 1995, 932). Taking $z' = [1,\ dA/dC]$ shows that $H^0_{KK} < 0$.

For zero pollution H^0 reads $U\{C(K), 0\} + \lambda[\phi(K) - \delta K - C(K) - A(C(K))]$. So H^0_K equals $\lambda(\phi_K - \delta) + dC/dK(U_C - \lambda + \lambda P_C/P_A)$ where (6.12) has been inserted. On the optimal path $U_C - \lambda + \lambda P_C/P_A = 0$. Hence, using (6.3) we find

$$H^0_{KK} = \lambda\phi_{KK} + \left(\frac{dC}{dK}\right)^2\left(U_{CC} - \lambda\,\frac{P_{CC} - 2P_{AC}P_C/P_A + P_{AA}P_C^2/P_A^2}{-P_A}\right)$$

which is negative due to $U_{CC} \leq 0$ and (6.3).

Note that H^0 is also concave at $\tilde{C}$ and $C_{P=0}$ because H^0_K is always $\lambda(\phi_K - \delta)$. Thus the slope of H^0 does not change at $\tilde{C}$ and $C_{P=0}$ while the curvature changes there, but remains concave. □

Proof of lemma 6.3

Remember that the assumptions about MRS, MRS_C, and MRS_P in (4.1) and (4.2) are ordinal and thus robust to monotonic transformations of the felicity function. Hence, a monotonic transformation does not change $\tilde{C}$, $C_{P=0}$ (see (6.8)), the $\dot{K} = 0$-line (see (6.13), (6.11), and (6.12)), and the $\dot{C} = 0$-line (see (6.14), (6.15), and (6.17)). The value of $\dot{K}$ at some given point in the phase diagram is also robust. The level of $\dot{C}$, in contrast, changes with the elasticity of intertemporal substitution because the fractions in (6.14), (6.15), and (6.17) are altered by a monotonic transformation of the felicity function: Consider for example the monotonic transformation $\tilde{U} = U^x$ where $0 < x < 1$ and $U > 0$ (if $U < 0$, we use $a > 0$ such that $U + a > 0$ and define $\tilde{U} = (U + a)^x)$. The whole proof remains valid if we instead consider $\tilde{U} = -U^x$ where $x < 0$. Since this transformation makes the differences between felicity levels attained at different points in time smaller, the elasticity of intertemporal substitution falls and individuals prefer a smooth development of felicity levels along the optimal path. We now show that for this transformation (6.2) remains valid and $\dot{C}$ changes such that the optimal path in the phase diagram shifts upwards. The transformation $\tilde{U} = U^x$ yields

$$\begin{array}{ll} \tilde{U}_C = xU^{x-1}U_C & \tilde{U}_{CC} = xU^{x-1}U_{CC} + x(x-1)U^{x-2}U_C^2 \\ \tilde{U}_P = xU^{x-1}U_P & \tilde{U}_{PP} = xU^{x-1}U_{PP} + x(x-1)U^{x-2}U_P^2 \\ & \tilde{U}_{CP} = xU^{x-1}U_{CP} + x(x-1)U^{x-2}U_CU_P. \end{array} \tag{6.18}$$

Moreover, using (6.1) we derive

$$\begin{aligned} \tilde{H}_{CC} &= xU^{x-1}H_{CC} + x(x-1)U^{x-2}(U_C^2 + 2U_CU_PP_C + U_P^2P_C^2) \\ \tilde{H}_{AA} &= xU^{x-1}H_{AA} + x(x-1)U^{x-2}U_P^2P_A^2 \\ \tilde{H}_{AC} &= xU^{x-1}H_{AC} + x(x-1)U^{x-2}(U_CU_PP_A + U_P^2P_AP_C). \end{aligned}$$

For the second equality sign of the next derivation we use $U_C = U_PP_A - U_PP_C$ which is (6.8) at an interior solution. We are allowed to use this equation because $H_{AA}H_{CC} - H_{AC}^2$ is only used for the interior solution as the previous proofs have shown. So we find

$$\begin{aligned} (\tilde{H}_{AA}\tilde{H}_{CC} - \tilde{H}_{AC}^2)/(xU^{x-1})^2 &= \big(H_{AA} + U_P^2P_A^2(x-1)/U\big) \\ &\quad \cdot \big(H_{CC} + (U_C + U_PP_C)^2(x-1)/U\big) - \big(H_{AC} + (U_CU_PP_A + U_P^2P_AP_C)(x-1)/U\big)^2 \\ &= \big(H_{AA} + U_P^2P_A^2(x-1)/U\big)\big(H_{CC} + U_P^2P_A^2(x-1)/U\big) - \big(H_{AC} + U_P^2P_A^2(x-1)/U\big)^2 \\ &= H_{AA}H_{CC} - H_{AC}^2 + U_P^2P_A^2(H_{AA} - H_{AC} + H_{CC} - H_{AC})(x-1)/U. \end{aligned} \tag{6.19}$$

Thus the transformed felicity function $\tilde{U}$ satisfies (6.2) if (6.2) holds before the transformation, i.e. $\tilde{H}_{CC} < 0$ and $\tilde{H}_{AA}\tilde{H}_{CC} - \tilde{H}_{AC}^2 > 0$ (see (6.19) and (6.9)) if $H_{CC} < 0$ and $H_{AA}H_{CC} - H_{AC}^2 > 0$ (for $x > 1$ this is not necessarily the case). Similarly, for $U_{CC} \leq 0$ we find $\tilde{U}_{CC} < 0$.

It remains to be shown that $\dot{C}$ changes. The term showing up in (6.17) becomes

$$\frac{\tilde{U}_{CC}}{\tilde{U}_C} = \frac{xU^{x-1}U_{CC}+x(x-1)U^{x-2}U_C^2}{xU^{x-1}U_C} = \frac{U_{CC}}{U_C} + \frac{(x-1)U_C}{U} < \frac{U_{CC}}{U_C}$$

and the fraction in (6.14) changes to

$$\frac{\tilde{U}_C + \tilde{U}_P P_C}{\tilde{H}_{CC}} = \frac{U_C + U_P P_C}{H_{CC} + (U_C + U_P P_C)^2(x-1)/U} > \frac{U_C + U_P P_C}{H_{CC}}$$

where the inequality sign follows from $H_{CC} < 0$ (see (6.2)). Finally, using (6.19) and (6.9) we can write the fraction in (6.15) as $P_A(MRS_P P_A + P_{AA} - P_{AC})$ multiplied by

$$\begin{aligned} \frac{\tilde{U}_P^2}{\tilde{H}_{AA}\tilde{H}_{CC}-\tilde{H}_{AC}^2} &= \frac{U_P^2}{H_{AA}H_{CC}-H_{AC}^2+U_P^2P_A^2(H_{AA}+H_{CC}-2H_{AC})(x-1)/U} \\ &< \frac{U_P^2}{H_{AA}H_{CC}-H_{AC}^2}. \end{aligned}$$

Therefore the absolute values of the fractions in (6.14), (6.15), and (6.17) all decrease. Thus this transformation makes $\dot{C}$ smaller to the left of the $\dot{C} = 0$-line. Hence, we must start at a higher value of C at K_0 to reach the steady state: The optimal path shifts upwards – for example, from the lowest path in figure 6.1 to the middle or highest path in figure 6.1 □

7. The Elasticity of Intertemporal Substitution: Reconsidering the Model of Selden and Song

7.1 Introduction

In chapters 5 and 6 we have considered a static and a dynamic version of a model in which pollution is generated by consumption. For some kinds of pollution this is a reasonable assumption. For example, exhaust components are generated by the consumption of car kilometres. Moreover, consumption often results in waste. Similarly, the consumption of washing powder leads to waste water. Furthermore, energy – often gained from fossil fuels – is needed for the consumption of services from household appliances such as fridges, stoves, washing machines, lawnmowers, and many more. Other kinds of pollution, however, are generated in the production process. For instance, this is the case when energy is used in the production process or when (toxic) intermediate or joint products – for example from the chemical industry – are released into the environment. So the often cited model of *Selden* and *Song* (1995) in which pollution depends on production capital might give some further insights into the PIR and is therefore worth analysing.

We extend the model of *Selden* and *Song* (1995) in this chapter.[1] First, we check which results are robust to changes in the elasticity of intertemporal substitution using the method developed in section 4.3. Hence, as shown in section 4.3 we must allow the cross-derivative of the felicity function to be nonzero. This enables us to show that – in contrast to the model of chapter 6 –

[1]Their model is based on *Forster* (1973). *Pezzey* (1989, 25–29) considers a simplified static version of the model.

the elasticity of intertemporal substitution has a large influence on the PIR: If the PIR in the model of *Selden* and *Song* (1995) is an EKC, we can decrease the elasticity of intertemporal substitution to obtain a monotonically rising or an N-shaped PIR. Intuitively, when the elasticity of intertemporal substitution is small, a smooth development path of felicity levels is preferred. Hence, we do not accept the high pollution levels around the turning point of the EKC. This would allow fast growth and thus would enable us to reach high felicity levels early on. Instead we prefer slower growth and a smooth, possibly monotonically rising development of pollution.

A second extension of *Selden* and *Song's* (1995) model is that we allow the cross-derivative of the pollution function to be nonzero as we did in chapters 5 and 6. Third, we restrict pollution to be nonnegative. Only in this extended model we can find conditions which ensure that the PIR is an EKC independent of the elasticity of intertemporal substitution. The level of pollution and of income at the turning point of the EKC, however, both rise with the elasticity of intertemporal substitution. In contrast, the steady state is again robust to changes in the elasticity of intertemporal substitution. We find that the steady state pollution level is zero under almost the same conditions as in chapter 6.

This chapter is organized as follows. In section 7.2 we present the assumptions of the model which are similar to the assumptions in chapter 6. In section 7.3 we derive the phase diagram. In section 7.4 we study the PIR and we find that the elasticity of intertemporal substitution is a crucial determinant of the PIR. The steady state pollution level is analysed in section 7.5. Section 7.6 concludes.

7.2 Additional assumptions

We again use all the assumptions made in section 5.2 with two exceptions. First, as in the model of chapter 6 we replace the static allocation of income $Y = C + A$ by

$$\dot{K} = \phi(K) - \delta K - C - A$$

where the variables are defined as in chapter 6 and where as before $\phi_K > 0$ and $\phi_{KK} < 0$. The initial capital stock K_0 is again assumed to be small because this is the interesting case in which an EKC might emerge.

Second, pollution $P \geq 0$ is no longer generated by consumption, but now results from capital.[2] All previous assumptions about $P(C, A)$ are now valid for $P(K, A)$: We assume that $P_K > 0$, $P_{KK} \geq 0$, $P_A < 0$, $P_{AA} > 0$, $P_{AK} = P_{KA} \leq 0$, and $\lim_{A \to 0} |P_A| < \infty$. *Selden* and *Song* (1995) assume $P_{AK} = 0$ which is less general (see section 5.2). The slope of the iso-pollution lines in figure 7.1 is given by $dA/dK = -P_K/P_A$ (see section 5.2). Thus condition (5.2) now reads

$$\left.\frac{dP_A}{dK}\right|_{P=\text{constant}} = P_{AK} + P_{AA} \left.\frac{dA}{dK}\right|_{P=\text{constant}} = \frac{-P_{AK}P_A + P_{AA}P_K}{-P_A} > 0. \tag{7.1}$$

Remember that condition (7.1) states that abatement becomes more expensive ($|P_A|$ falls) when we move along the iso-pollution lines to the right. This condition might or might not hold. It holds

[2] Pollution might also result from output $\phi(K)$ instead of being caused by capital K. If pollution is rising with K, it is also rising with $\phi(K)$ because $\phi(K)$ is strictly increasing in K. Thus the sign of the slope of the PIR is the same for both models. Following the literature (*Forster* 1973, *Gradus* and *Smulders* 1993, *Selden* and *Song* 1995, *van Ewijk* and *van Wijnbergen* 1995, *Smulders* and *Gradus* 1996, and *Ansuategi* and *Perrings* 2000) we choose pollution to be generated by capital.

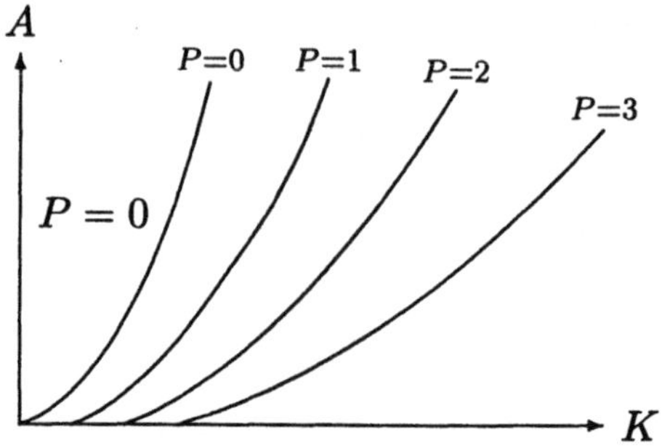

Figure 7.1: The iso-pollution lines.

for standard functional forms (see appendix A on page 257) or for $P_{AK} = 0$. Condition (5.1) is no longer important in this model.

Additionally, we need two assumptions to assure that the model is well-behaved, i.e. to exclude multiple equilibria and to ensure that the second order conditions for a maximum are satisfied, that consumption never changes discontinuously, and that the steady state is a saddle point. First, we must assume that $P(K, A)$ is convex. Then the Hessian $D^2P = \begin{bmatrix} P_{KK} & P_{AK} \\ P_{AK} & P_{AA} \end{bmatrix}$ is positive semidefinite. Therefore, on the one hand,

$$P_{AA}P_{KK} - P_{AK}^2 \geq 0 \tag{7.2}$$

and, on the other hand, $z'[D^2P]z \geq 0$ for all vectors z (*Mas-Colell et al.* 1995, 936 and 932). Taking $z' = [1, \ -P_K/P_A]$ yields

$$P_{KK} - 2P_{AK}P_K/P_A + P_{AA}P_K^2/P_A^2 \geq 0. \tag{7.3}$$

As shown in (6.3) this is exactly the condition that the iso-pollution lines are convex as drawn in figure 7.1. Thus if $P(K, A)$ is convex, the iso-pollution lines are also convex. In appendix A (see page 257) we show that two standard functional forms satisfy (7.2) for all or for many parameter constellations. If we assumed $P_{AK} = 0$, as is common in the literature, we would implicitly assume that (7.2) and thus (7.3) both hold.

Second, as in chapter 6 we have to assume that the Hessian $D^2H = \begin{bmatrix} H_{CC} & H_{AC} & H_{CK} \\ H_{AC} & H_{AA} & H_{AK} \\ H_{CK} & H_{AK} & H_{KK} \end{bmatrix}$ is negative semidefinite where[3]

$$\begin{array}{ll} H_{CC} = U_{CC} & H_{AA} = U_{PP}P_A^2 + U_P P_{AA} \\ H_{AC} = U_{CP}P_A & H_{KK} = U_{PP}P_K^2 + U_P P_{KK} + \lambda\phi_{KK} \\ H_{CK} = U_{CP}P_K & H_{AK} = U_{PP}P_A P_K + U_P P_{AK}. \end{array} \tag{7.4}$$

It follows that $H_{CC} \leq 0$ and $H_{AA}H_{CC} - H_{AC}^2 \geq 0$. However, we need the slightly stronger assumption that

$$H_{CC} = U_{CC} < 0 \quad \text{and} \quad H_{AA}H_{CC} - H_{AC}^2 > 0. \tag{7.5}$$

As in chapter 6 these are mainly constraints on the permissible values of U_{CC}, U_{PP}, and U_{CP}. Put differently, these are constraints on the permissible monotonic transformations of the felicity function when studying whether or not the results are robust to changes in the elasticity of intertemporal substitution.

Thus we do not restrict U_{CP} to equal zero as *Selden* and *Song* (1995) do. Instead, we keep assuming that consumption and environmental quality in terms of reducing pollution ($-P$) are both normal goods, i.e. $MRS_C < 0$ and $MRS_P < 0$ (see (4.1) and (4.2)). As mentioned in subsection 4.3.1 assuming $U_{CP} = 0$ means that we cannot check whether the results are robust to nonlinear monotonic transformations of the felicity function.[4]

[3]In the model of chapter 6 D^2H is also a 3×3 matrix. Since $H_{CK} = H_{AK} = 0$ and $H_{KK} = \lambda\phi_{KK} < 0$ in chapter 6, however, we simplified there.

[4]Furthermore, if we assumed $U_{CP} = 0$ in this model, we would be led to conclude that the $\dot{C} = 0$-line is robust although it is not (see (7.26) and (7.27) below).

7.3 Derivation of the phase diagram

The maximization problem is the same as in chapter 6 except for $P(C, A)$ which is replaced by $P(K, A)$. Thus pollution is again a flow. The current value Lagrangian is

$$L = U(C, P(K, A)) + \lambda(\phi(K) - \delta K - C - A) + \gamma A + \mu P(K, A).$$

The first order conditions read

$$\begin{aligned} L_C &= U_C - \lambda = 0 & (7.6) \\ L_A &= (U_P + \mu)P_A - \lambda + \gamma = 0 & (7.7) \\ -L_K &= -(U_P + \mu)P_K - \lambda(\phi_K - \delta) & \\ &= \lambda(\delta - \phi_K - \frac{(U_P + \mu)P_K}{(U_P + \mu)P_A + \gamma}) = \dot{\lambda} - \rho\lambda, & (7.8) \end{aligned}$$

where (7.7) is used in (7.8). The transversality condition is

$$\lim_{t \to \infty} e^{-\rho t} \lambda(t)[K(t) - \check{K}(t)] \geq 0 \qquad (7.9)$$

where $\check{K}(t)$ is capital on the optimal path and $K(t)$ is capital on any other feasible path. Combining (7.6) and (7.7) yields

$$MRS = -\frac{U_C}{U_P} = -P_A \frac{U_P + \mu}{U_P} - \frac{\gamma}{U_P} \qquad (7.10)$$

or $MRS = -P_A$ at an interior solution where $\gamma = \mu = 0$. (7.10) is similar to (6.8) which is identical to (5.6), but the P_C-term in (6.8) or (5.6) drops out in (7.10) as consumption is no longer pollution-enhancing. Hence, the results of chapter 5 no longer apply in this model. In fact, the pollution-consumption relationship can no longer be derived as easily as in chapter 6. Therefore and because the pollution-consumption relationship is only an intermediate result, we will not calculate it in this model.

We now derive the phase diagram. As in chapter 6 the phase diagram is three-dimensional with consumption, abatement expenditures, and capital on the three axes. However, for each given

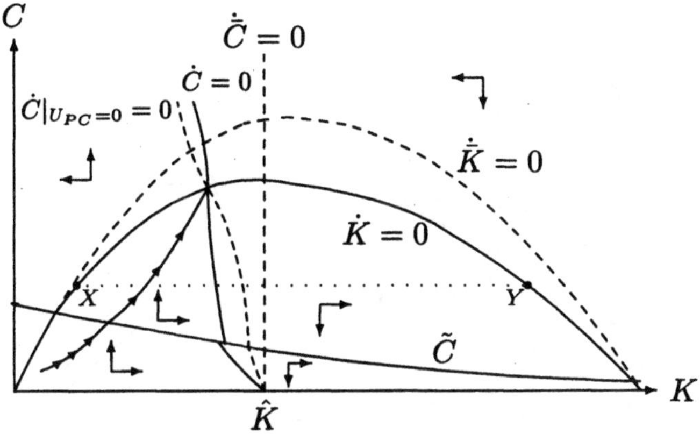

Figure 7.2: Phase diagram if (7.1) holds and $U_{CP} < 0$.

level of capital and consumption we can choose the optimal level of abatement according to (7.10) and thus make the phase diagram two-dimensional. We first derive the $\tilde{C}$-line below which abatement expenditures are optimally chosen to be zero and the $P = 0$-line above which pollution is zero. Note that the $P = 0$-line does not exist if zero pollution is technically not feasible. Under the assumptions made in section 7.2 we obtain the following lemma.

Lemma 7.1 *1. The* $\tilde{C}$*-line is downward sloping as shown in figures 7.2–7.4.*

2. As shown in figure 7.3 (7.4) the $P = 0$*-line is upward (downward) sloping, if (7.1) is satisfied (violated), i.e. if abatement becomes more (less) expensive with rising capital while holding pollution constant. The* $P = 0$*-line always lies above the* $\tilde{C}$*-line.*

The proof is given in the appendix. The $\tilde{C}$-line is not horizontal as in chapter 6. Instead the $\tilde{C}$-line is falling: The higher capital, the higher is pollution and the more attractive is abatement such that we start to increase abatement at lower levels of consumption.

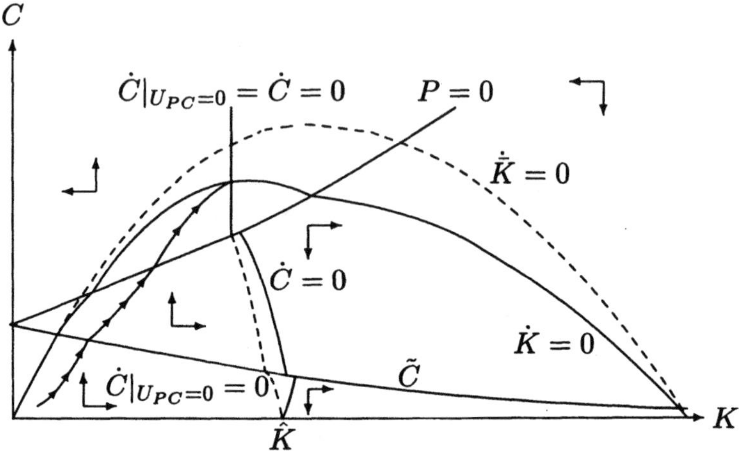

Figure 7.3: Phase diagram with zero-pollution steady state if (7.1) holds and $U_{CP} > 0$.

The sign of the slope of the $P = 0$-line depends on (7.1) because $MRS(C, P) = -P_A$ must hold at an interior solution (see (7.10)). If capital rises, if pollution is held constant at zero, and if (7.1) is satisfied (violated), $|P_A|$ decreases (increases) along the $P = 0$-line. Thus the MRS must also fall (rise) which necessitates an increase (decrease) in consumption.

Next, we derive the $\dot{K} = 0$- and the $\dot{C} = 0$-line. Since the $\dot{C} = 0$-line is difficult to calculate, we first set $U_{CP} = 0$ to derive the $\dot{C}|_{U_{CP}=0} = 0$-line which serves as a reference for the general $\dot{C} = 0$-line with $U_{CP} \neq 0$. The $\dot{\tilde{K}} = 0$- and the $\dot{\tilde{C}} = 0$-line for the standard Ramsey model without pollution ($P = P_K = P_A = 0$) are identical to the $\dot{\tilde{K}} = 0$- and the $\dot{\tilde{C}} = 0$-line of the model of chapter 6 since these models only differ in how pollution is generated. Thus the $\dot{\tilde{C}} = 0$-line is vertical at $\hat{K}$ where $\phi_K(\hat{K}) = \rho + \delta$ (see lemma 6.1). All these lines are shown in figures 7.2–7.4.

Lemma 7.2 *1. The maximum of the $\dot{K} = 0$-line lies to the left of the maximum of the $\dot{\tilde{K}} = 0$-line. Below the $\tilde{C}$-line the $\dot{K} = 0$- and the $\dot{\tilde{K}} = 0$-line are identical. When the $\dot{K} = 0$-line crosses the $\tilde{C}$-line ($P = 0$-line) from below, it becomes suddenly flatter (steeper). Below (above) the $\dot{K} = 0$-line capital rises (falls).*

2. The $\dot{C}|_{U_{CP}=0} = 0$-line starts at point $(\hat{K}, 0)$ in the phase diagram. Below the $\tilde{C}$-line it is downward sloping. At an interior solution the $\dot{C}|_{U_{CP}=0} = 0$-line is downward (upward) sloping if (7.1) is satisfied (violated). Above the $P = 0$-line the $\dot{C}|_{U_{CP}=0} = 0$-line is vertical. The $\dot{C}|_{U_{PC}=0} = 0$-line always lies to the left of the $\dot{\tilde{C}} = 0$-line.

3. The $\dot{C} = 0$-line is identical to the $\dot{C}|_{U_{CP}=0} = 0$-line at point $(\hat{K}, 0)$, at the steady state, and above the $P = 0$-line. To the left (right) of the $\dot{C} = 0$-line consumption rises (falls). If $U_{CP} < 0$, the $\dot{C} = 0$-line lies to the left of the $\dot{C}|_{U_{CP}=0} = 0$-line below the $\tilde{C}$-line. At an interior solution and below the $\dot{K} = 0$-line the $\dot{C} = 0$-line for $U_{CP} < 0$ lies to the left (right) of the $\dot{C}|_{U_{CP}=0} = 0$-line if (7.1) is satisfied (violated). Above the $\dot{K} = 0$-line or for $U_{CP} > 0$ it is the other way round. The $\dot{C} = 0$-line is discontinuous when it crosses the $\tilde{C}$- and the $P = 0$-line.

The proof is provided in the appendix. As in chapter 6 the vertical distance between the $\dot{\tilde{K}} = 0$- and the $\dot{K} = 0$-line are the abatement expenditures. This again explains the kinks of the $\dot{K} = 0$-line when it crosses the $\tilde{C}$- and the $P = 0$-line (cf. page 145). There is exactly one steady state lying at the intersection of the $\dot{K} = 0$- and the $\dot{C}|_{U_{CP}=0} = 0$-line which is identical to the $\dot{C} = 0$-line at the steady state. Therefore it is no problem for the existence of a steady state that the $\dot{C} = 0$-line is discontinuous when it crosses the $\tilde{C}$ and the $P = 0$-line. The $\dot{C} = 0$-line is discontinuous because the difference between $\dot{C} = 0$- and the $\dot{C}|_{U_{CP}=0} = 0$-line below the

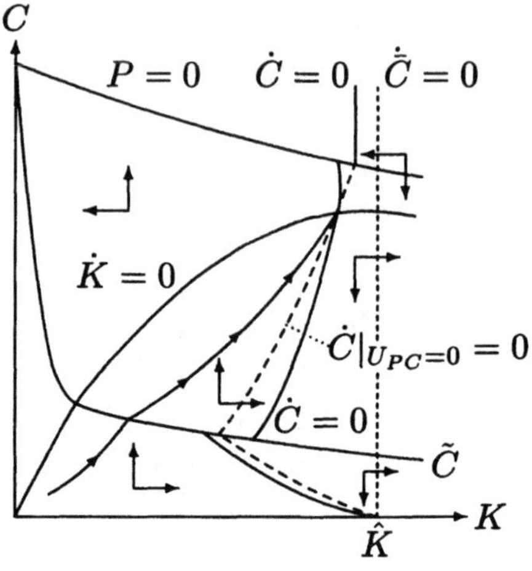

Figure 7.4: Phase diagram if (7.1) is violated and $U_{CP} < 0$.

$\tilde{C}$-line is not equal to this difference at an interior solution and the difference is zero above the $P = 0$-line. If (7.1) is violated, the steady state lies at a higher capital level because then abatement becomes cheaper with rising capital (holding pollution constant). Without pollution the steady state would lie at the intersection of the $\dot{K} = 0$- and the $\dot{\tilde{C}} = 0$-line. With pollution the steady state capital stock is smaller because capital is polluting and the steady state consumption level is smaller because part of income is used for abatement instead of consumption and because with the lower capital stock income, $Y = \phi(K) - \delta K$ $(= C + A$ at the steady state), is lower. *Tahvonen* and *Kuuluvainen* (1993, 106) achieve a similar result with a stock instead of a flow pollutant.

7.4 The pollution-income relationship

Before we can analyse the PIR, we have to derive the optimal path of the economy. The following lemma – which is the analogue to lemma 6.2 – collects a few important results.

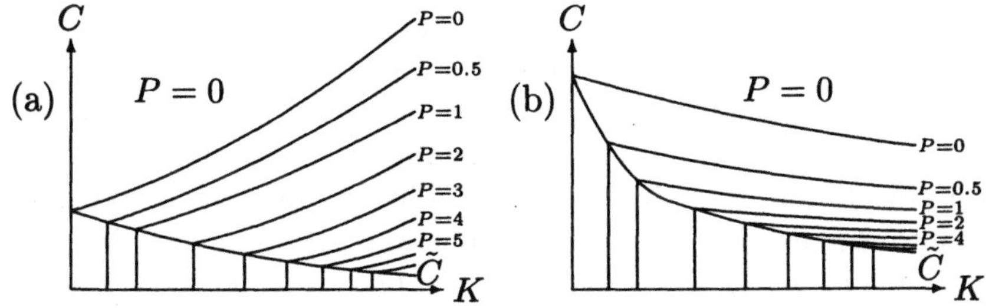

Figure 7.5: The constant-pollution lines if (7.1) is (a) satisfied or (b) violated.

Lemma 7.3 *1. The steady state is a saddle point in all the cases we consider.*

2. Discontinuous changes in consumption are never optimal.

3. The path into the saddle point is the only path that satisfies both the first order conditions (7.6) – (7.8) and the transversality condition (7.9).

4. The second order conditions for a maximum are satisfied.

The proof is relegated to the appendix. We can also see from figures 7.2–7.4 that the steady state is a saddle point. Theoretically possible sudden changes in the control variable are not optimal. The saddle path gives maximal, not minimal utility.

To analyse the PIR we can again consider the pollution-capital relationship as in chapter 6: Income $\phi(K) - \delta K$ (or GDP $= \phi(K)$) increases with capital along the optimal path since the steady state lies to the left of the $\dot{\tilde{C}} = 0$-line lying at $\hat{K}$ (see lemma 7.2.2) and since income reaches its maximum level at $K^* > \hat{K}$ (see figure 6.1 on page 145).

We need one last tool to be able to derive the PIR: The lines in the phase diagram on which pollution stays constant are called the constant-pollution lines. They are shown in figure 7.5 and should not be confused with the iso-pollution lines in figure 7.1. As long as abatement is zero, pollution depends only on capital, $P(K, 0)$. Thus below the $\tilde{C}$-line the constant-pollution lines are vertical. Above the $\tilde{C}$-line the constant-pollution lines behave as the $P = 0$-line does since the $P = 0$-line is one of the constant-pollution lines (the slope of the constant-pollution lines is given by (7.16) in the proof of lemma 7.1): If (7.1) is satisfied (violated), the constant-pollution lines are upward (downward) sloping as in figure 7.5a (7.5b). In contrast, in chapter 6 the constant-pollution lines are horizontal because in chapter 6 pollution depends only on consumption (which also determines abatement). Here, however, pollution depends also on capital and the constant-pollution lines are therefore no longer horizontal.

We can now immediately derive the PIR. If the path of the economy is more positively sloped than the constant-pollution lines in figure 7.5, pollution is falling and if the path is more negatively sloped, pollution is rising.[5] Thus the PIR can easily be derived by drawing the constant-pollution lines of figure 7.5 into figures 7.2–7.4. An example is shown in figure 7.6. However, as in the model of chapter 6 the optimal path is not robust to a monotonic transformation of the felicity function:

Lemma 7.4 *In the phase diagram the optimal path and the $\dot{C} = 0$-line below the $P = 0$-line change with the elasticity of intertemporal substitution while the constant-pollution lines (including the $P = 0$-line), the $\tilde{C}$-, the $\dot{K} = 0$-, the $\dot{C}|_{U_{CP}=0} = 0$-line, and thus the steady state are all robust to changes in the elasticity of intertemporal substitution.*

---

[5]This can also be derived mathematically: Inserting (7.13) and (7.14) from the appendix into $\dot{P} = P_K\dot{K} + P_A\dot{A}$ we derive $\dot{P} = \{-MRS_C P_A \dot{C} + (P_{AA}P_K - P_{AK}P_A)\dot{K}\}/\{MRS_P P_A + P_{AA}\}$. Thus pollution falls, i.e. $\dot{P} < 0$, if $\dot{C}/\dot{K} > (P_{AA}P_K - P_{AK}P_A)/(MRS_C P_A)$ where the left hand side is the slope of the optimal path and the right hand side equals the slope of the constant-pollution lines (see (7.16)).

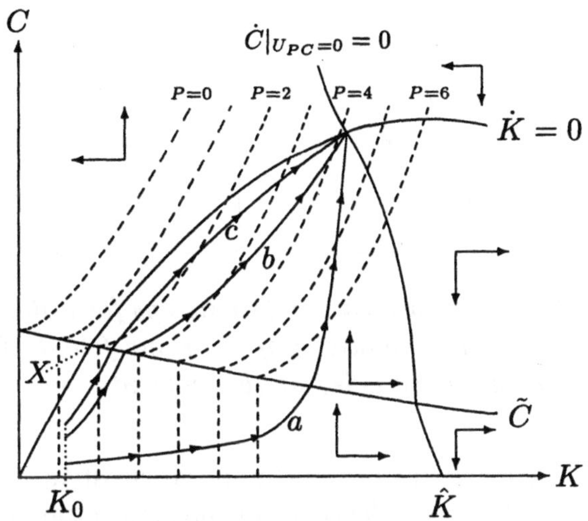

Figure 7.6: Paths for different monotonic transformations of the same felicity function.

The proof is given in the appendix. Three paths for three different monotonic transformations of the same ordinal felicity function are shown in figure 7.6.[6] That capital and consumption grow along the optimal path is robust to changes in the elasticity of intertemporal substitution, but the curvature of the path is not robust. From figure 7.6 we can immediately see the crucial result of this chapter: On path a pollution rises as long as $A = 0$, but falls as soon as $A > 0$ because the path is more positively sloped than the constant-pollution lines. An EKC arises (see path a in figure 7.7a). In contrast, on path b in figure 7.6 and 7.7a the PIR is monotonically rising. On path c in figure 7.6 and 7.7a pollution first rises, then falls, and finally rises again: An N-shaped PIR emerges. Hence, we conclude that the shape of the PIR – i.e. whether it is monotonically rising, an EKC, or N-shaped –

[6]The three different $\dot{C} = 0$-lines, however, are not shown as this would make the figure overladen.

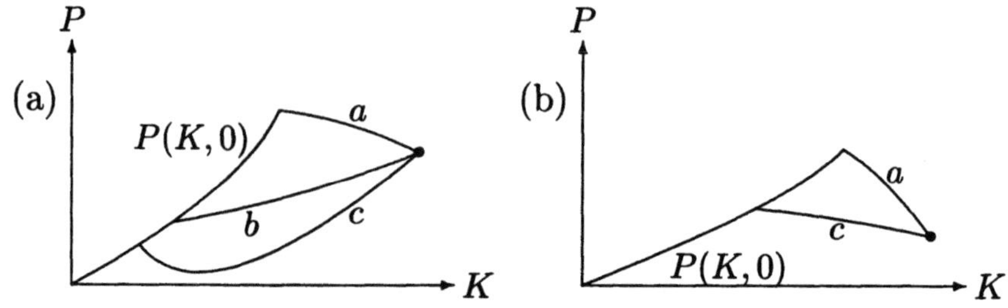

Figure 7.7: The pollution-capital relationships for different monotonic transformations of the same felicity function if (7.1) is (a) satisfied or (b) violated.

changes with the elasticity of intertemporal substitution. In fact, an EKC exists only if the elasticity of intertemporal substitution is sufficiently high. This is stated in the following proposition together with two other results.

Proposition 7.1 *1. If the steady state lies below the $\tilde{C}$-line, the PIR is monotonically increasing and robust to changes in the elasticity of intertemporal substitution.*

2. Suppose that there is an interior steady state, that (7.1) holds (i.e. that abatement becomes more expensive with rising capital along the constant-pollution lines), that the steady state pollution level (P_{SS}) is higher than the pollution level $\tilde{P}$ at the intersection of the $\dot{K} = 0$- and the $\tilde{C}$-line (i.e. at point X in figure 7.6), and that the constant-pollution line through the steady state lies always below the $\dot{K} = 0$-line for capital stocks smaller than in the steady state. Then if the PIR is an EKC, we can decrease the elasticity of intertemporal substitution such that the PIR becomes monotonically rising or N-shaped (see figure 7.7a).

3. Suppose that (7.1) is violated or that the steady state lies above the $\dot{P} = 0$-line. Then the PIR is an EKC. The pollution and income level at the turning point of the EKC, *however, are both high when the elasticity of intertemporal substitution is large (see figure 7.7b).*

Proof. 1. Below the $\tilde{C}$-line the constant-pollution lines are vertical. Thus pollution depends only on capital. So whether the optimal path starts at a low or high consumption level has no effect on the PIR. This holds also true if the steady state lies above the $\tilde{C}$-line as long as abatement expenditures are zero along the optimal path: The pollution-capital relationship follows the $P(K,0)$-line in figure 7.7.

2. This is the case of figure 7.6: The steady state is interior and the constant-pollution lines are rising since (7.1) is satisfied. The other two preconditions are necessary: First, if P_{SS} was smaller than $\tilde{P}$, every PIR could be an EKC: Every path which is optimal for some monotonic transformation of the felicity function crosses the $\tilde{C}$-line at a higher pollution level than $\tilde{P}$ and would then fall back to P_{SS}. Second, if the constant-pollution line through the steady state lay partly above the $\dot{K} = 0$-line for capital stocks smaller than in the steady state, every path would reach a higher pollution level than P_{SS} somewhere along the path.[7] In the appendix we show that by lowering the elasticity of intertemporal substitution (such that individuals prefer to smooth felicity levels

[7]The constant-pollution line through the steady state lies always below the $\dot{K} = 0$-line if the constant-pollution lines are convex (and if $P_{SS} > \tilde{P}$). This is possible as a simple example shows: For simplicity assume $P_{AK} = 0$. Then the slope of the constant-pollution lines is given by $B := (dC/dK)|_{P=constant} = P_{AA}P_K/(MRS_C P_A)$ (see (7.16)). Hence, the constant-pollution lines are convex as we derive

$$\left.\frac{d^2C}{dK^2}\right|_{P=constant} = \frac{dB}{dK} + \frac{dB}{dC}B$$
$$= \frac{MRS_C P_A (P_{AAA} P_K \partial A/\partial K + P_{AA} P_{KK}) - P_{AA} P_K MRS_C P_{AA} \partial A/\partial K}{MRS_C^2 P_A^2}$$
$$+ \frac{MRS_C P_A P_{AAA} P_K \partial A/\partial C - P_{AA} P_K (MRS_{CC} P_A + MRS_C P_{AA} \partial A/\partial C)}{MRS_C^2 P_A^2} \frac{P_{AA} P_K}{MRS_C P_A}$$

which is positive unless $P_{AAA} \ll 0$ and $MRS_{CC} \ll 0$ ($\partial A/\partial K > 0$ and $\partial A/\partial C > 0$ are given in (7.15)).

along the optimal path), the optimal path shifts upwards as in the model of chapter 6. In extreme cases the optimal path approaches the $\dot{K} = 0$-line. Under our assumptions the PIR of a path close to the $\dot{K} = 0$-line is monotonically rising or N-shaped depending on the exact shape of the constant-pollution lines (relative to the $\dot{K} = 0$-line). Hence, when the PIR is an EKC, we can transform the felicity function monotonically such that the PIR becomes monotonically rising or N-shaped. As shown in the appendix we can transform the felicity function such that the elasticity of intertemporal substitution falls while (7.5) remains satisfied and D^2H as defined on page 167 remains negative semidefinite. However, when we increase the elasticity, these conditions might no longer be satisfied. Therefore the statement might be wrong that every monotonically rising PIR can be transformed in an EKC by increasing the elasticity of intertemporal substitution.

3. If (7.1) is violated the constant-pollution lines are downward sloping above the $\tilde{C}$-line, while the optimal path is always upward sloping. So there is an EKC: Pollution rises below the $\tilde{C}$-line, but falls above it. As shown in the appendix (see proof of proposition 7.1.2) a large (small) elasticity of intertemporal substitution implies that the optimal path stays below the $\tilde{C}$-line until a high (low) capital stock is reached (cf. path a (c) in figure 7.6 where the constant-pollution lines must now be imagined to be downward sloping). Thus the pollution and income level at the turning point of the EKC are both high (low) as shown by path a (c) in figure 7.7b. If pollution is zero at the steady state, pollution rises as long as there is no abatement and must later fall back to zero. Again the turning point of the EKC is not robust to monotonic transformations of the felicity function. □

Let us compare these results to those of the model of chapter 6 in which pollution is generated by consumption. For small income levels where abatement is zero the PIR is robust to a monotonic transformation of the felicity function in the present model. In the model of chapter 6, in contrast, the elasticity of intertemporal substitution determines the initial level of consumption and therefore the initial level of pollution. However, in the present model not only the income level at the turning point of the EKC – if there is one – rises with the elasticity of intertemporal substitution, but also the pollution level at the turning point. Even worse, in the model of chapter 6 the shape of the PIR is robust to changes in the elasticity of intertemporal substitution, but here it is not robust if the steady state is interior and if (7.1) holds. This, however, is the only case analysed by *Selden* and *Song* (1995) since they only consider an interior steady state and since they assume that (7.1) holds as $P_{AA} > 0$ and $P_{AK} = 0$ in their model.

In the extended model the shape of the PIR is robust if (7.1) is violated, if pollution becomes zero, or if abatement expenditures stay zero. Remember that (7.1) is fulfilled for two standard functional forms (see appendix A on page 257). Furthermore, for some pollutants zero pollution is technically not feasible. Hence, that case seems to be relevant in which the elasticity of intertemporal substitution is crucial for the shape of the PIR.

If (7.1) holds and if the steady state pollution level is sufficiently high, the PIR is an EKC only if the elasticity of intertemporal substitution is sufficiently high. Intuitively, if the elasticity of intertemporal substitution is small, we prefer a smooth development of felicity levels over time. Therefore we prefer to reach the steady state on a smooth, possibly monotonically rising pollution path. Furthermore, we start with a high consumption level C_0 and thus with a high felicity level (since $U(C_0, P(K_0, 0))$ where $P(K_0, 0)$ is given). In contrast, if the elasticity of intertemporal substitution is high, we start with a small consumption and felicity level and we willingly endure high pollution levels along the path in order to reach the higher felicity levels at the steady state early on. Whether or not high pollution levels are reached, is important if there are unexpected irreversible damages above some pollution level.

If the steady state pollution level is zero, however, an EKC emerges independent of the elasticity of intertemporal substitution. According to lemma 7.4 the steady state is robust to changes in the elasticity of intertemporal substitution. In the next section we will show that the steady state pollution level is determined by almost the same causes as in chapter 6. Hence, if these causes are sufficiently strong (such that $P_{SS} = 0$), they also produce an EKC in this model, but the turning point of the EKC is determined by the elasticity of intertemporal substitution: If the elasticity is small, the pollution level at the turning point is small since we prefer a smooth development. Accordingly, the income level at the turning point is also small. If the causes are not sufficiently strong, the elasticity of intertemporal substitution is also important for the shape of the PIR.

7.5 The steady state pollution level

To derive the factors which determine whether or not the steady state pollution level is zero we consider figure 7.8. This figure is the analogue to figure 5.2 on page 118. The concave lines are again the indifference curves. The ***consumption possibilities curve*** CPC for a given steady state capital stock K_{SS} is now defined by $P(K, A) = P(K_{SS}, Y_{SS} - C)$ where steady state income Y_{SS} is given by $\phi(K_{SS}) - \delta K_{SS}$ and is used for $C + A$ (in general, for $C + A + \dot{K}$, but $\dot{K} = 0$ at the steady state). If Y_{SS} would be used for consumption only, we would be at point A in figure 7.8, i.e. at point $(Y_{SS}, P(K_{SS}, 0))$. Point A is called the starting point of the CPC. The CPC is rising and convex since $dP/dC = -P_A > 0$ and $d^2P/dC^2 = P_{AA} > 0$. The slope of the CPC can again be denoted by the marginal rate of transformation, $MRT = -P_A$ (in chapters 5 and 6 the MRT was defined by $P_C - P_A$). The steady state is the point of the CPC giving the highest (ordinal) felicity level (point B with $P = 0$ in figure 7.8). For an interior solution

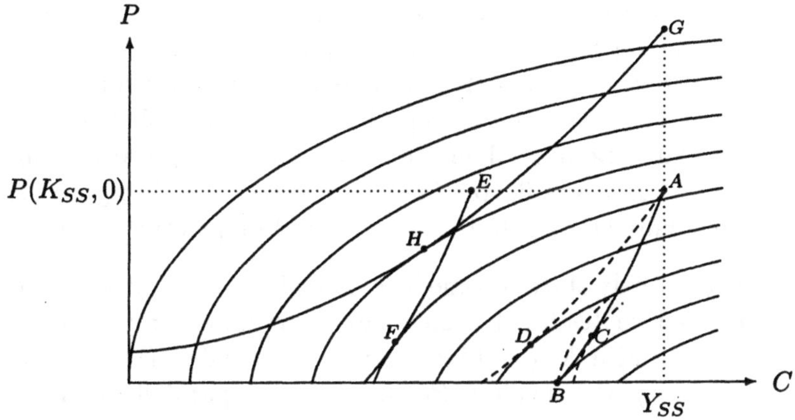

Figure 7.8: Indifference curves, different CPCs, and the corresponding steady states.

this means that the indifference curve is tangent to the CPC, i.e. that $MRS = -P_A = MRT$ (see (7.10)).

Pollution is zero at the steady state if on the C-axis in figure 7.8 the MRS is smaller than the MRT (see (7.10)). Thus steady state pollution is zero if the following conditions all hold simultaneously:

- If the indifference curves are flat and become quickly flatter with rising consumption, i.e. if the MRS is small and $|MRS_C|$ high. A tendency to satiation (U_C small and falling with rising C) or 'green' preferences ($|U_P|$ high and rising with C) cause the MRS to be small and become smaller. If the indifference curves were steeper (as the dashed indifference curves in figure 7.8), pollution would be positive at the steady state (point C).

- If abatement is cheap, i.e. if the $MRT = -P_A$ is high all along the CPC (implying that P_{AA} is small). This means that the CPC is steep and that the point of the CPC at zero pollution lies at a high consumption level (for a given starting point). Note that if (7.1) is satisfied (violated), $|P_A|$ falls (rises) with rising capital when pollution is held constant and thus a high steady state capital stock K_{SS} makes the CPC flatter (steeper). Comparing figures 7.3 and 7.4 we immediately see that zero pollution is more probable if (7.1) is violated because then the $P = 0$-line is downward sloping. If the CPC was flatter (as the dashed CPC in figure 7.8), the steady state pollution level would be positive (point D).

- If the steady state income level is high, i.e. if capital is highly productive ($\phi(K_{SS})$ high) and depreciates only slowly (δ small) and if K_{SS} is high. If Y_{SS} was smaller (holding K_{SS} constant), the whole CPC would move parallel to the left. The new starting point would be point E and the steady state point F in figure 7.8.

- If $P(K_{SS}, 0)$ is small, i.e. if capital is only slightly polluting and if K_{SS} is small. If production was more pollution-intensive, the starting point of the CPC would be higher (point G in figure 7.8) and steady state pollution positive (point H).

Note that the effect of the steady state capital stock K_{SS} is ambiguous. For the CPC starting at point G zero pollution is not feasible even if all income was used for abatement. To make zero pollution feasible the CPC would have to be steeper due to cheaper abatement, the steady state income level would have to be higher due to more productive capital, or capital would have to be less polluting.

For an interior steady state the analysis is exactly analogous: The steady state pollution level is small, if the indifference curves are flat due to a tendency to satiation or 'green' preferences, if the CPC is steep due to cheap abatement opportunities, if the steady state income level is high due to a highly productive production technology, and if capital is only slightly polluting (cf. figure 7.8).

Hence, pollution is zero (or small) under almost the same conditions as in chapter 6. Therefore these conclusions are valid independent of whether pollution is generated by consumption or by capital.

7.6 Conclusion

In this chapter we have found an additional important determinant of the PIR: The elasticity of intertemporal substitution. *Selden* and *Song* (1995) conclude that an EKC emerges from their model. This conclusion, however, is only correct when the elasticity of intertemporal substitution is sufficiently high. If the elasticity of intertemporal substitution is lowered, the PIR becomes monotonically rising or N-shaped. The intuition is that if the elasticity of intertemporal substitution is small, we prefer a smooth development of felicity levels. Therefore we let pollution develop smoothly, possibly monotonically, and we do not accept high pollution levels along the path which would result in an EKC. Hence, in the model of *Selden* and *Song* (1995) the shape of the PIR, i.e. whether it is monotonically rising, an EKC, or N-shaped, changes with a monotonic transformation of the felicity function.

We have not only reconsidered the model of *Selden* and *Song* (1995), but we have also extended it. First, we allowed the cross-derivative of the felicity function to be nonzero. This is necessary if we want to analyse which results are robust to changes in the elasticity of intertemporal substitution. To check for robustness we have again applied the method developed in section 4.3. Second, we restricted pollution to be nonnegative and we allowed the cross-derivative of the pollution function to be nonzero. The latter two extensions enabled us to prove that the PIR is an EKC, if the steady state pollution level is zero or if abatement becomes cheaper when we increase capital and abatement such that pollution stays constant. For standard functional forms, however, abatement becomes more expensive, not cheaper. Moreover, for certain kinds of pollution zero pollution is technically not feasible. Furthermore, the pollution and income level at the turning point of the EKC are both smaller when the elasticity of intertemporal substitution is small.

The steady state is robust to changes in the elasticity of intertemporal substitution. The steady state pollution level is zero, if the following four conditions all hold:

- There is a tendency to satiation in consumption or preferences are 'green'. Then consumption is unattractive so that a large fraction of income is spent on abatement.
- Abatement is cheap. This implies that only a low level of abatement expenditures is necessary to clean up a given amount of pollution.
- Capital is highly productive. This allows for a high income level which makes higher levels of consumption and abatement and thus lower levels of pollution possible.
- Capital causes only little pollution. Then for given levels of capital and of abatement expenditures the ensuing pollution level is lower.

These are almost the same conditions as in chapter 6. Therefore these conclusions are valid independent of whether pollution is generated by consumption or by capital. If these four conditions are satisfied, a high elasticity of intertemporal substitution causes the income and pollution level at the turning point of the EKC to be high. If they are not satisfied, the elasticity of intertemporal substitution is also crucial for the shape of the PIR: An EKC emerges only if the elasticity of intertemporal substitution is sufficiently high.

Appendix: Proofs

Proof of lemma 7.1

1. We first have to prove that abatement expenditures are zero at small consumption levels. Since $\lim_{C\to 0} MRS = \infty$, $\lim_{A\to 0} |P_A| < \infty$, and $(U_P + \mu)/U_P \leq 1$, the left hand side of (7.10) would be greater than the right hand side at small consumption levels if γ was zero. So γ is positive and A is zero. The $\tilde{C}$-line can be defined as the line on which abatement expenditures are optimally chosen to be zero even if $A < 0$ was feasible, i.e. it is defined by $MRS(C, P(K,0)) = -P_A(K,0)$ (see (7.10)). Holding $A = 0$ this leads to $MRS_C dC + MRS_P P_K dK = -P_{AK} dK$ or

$$\left.\frac{dC}{dK}\right|_{A=0} = -\frac{MRS_P P_K + P_{AK}}{MRS_C} < 0. \tag{7.11}$$

Thus the $\tilde{C}$-line is downward sloping.

2. Before we can analyse the $P = 0$-line, we have to derive how the optimal level of abatement changes with consumption and capital at an interior solution. Setting $\gamma = \mu = 0$ we derive from (7.6) – (7.8) that the second derivatives of the Hamiltonian H (L without $\gamma A + \mu P$) are given in (7.4). Thus differentiating $L_C = 0 = L_A$ with respect to time yields (assuming $\gamma = \mu = 0$ for L_A)

$$H_{CC}\dot{C} + H_{AC}\dot{A} + H_{CK}\dot{K} - \dot{\lambda} = 0 = H_{AA}\dot{A} + H_{AC}\dot{C} + H_{AK}\dot{K} - \dot{\lambda}. \tag{7.12}$$

For an interior solution we therefore obtain

$$\dot{A} = \frac{(H_{CC} - H_{AC})\dot{C} + (H_{CK} - H_{AK})\dot{K}}{H_{AA} - H_{AC}}. \tag{7.13}$$

At an interior solution we also find[8]

[8] First, using (4.2) and (7.4) and inserting $U_C/U_P = P_A$ which is (7.10) at an interior solution we find $MRS_P P_A + P_{AA} = (-U_{CP}U_P P_A + U_{PP}U_C P_A)/U_P^2 +$

$$
\begin{aligned}
(H_{AA} - H_{AC})/U_P &= MRS_P P_A + P_{AA} > 0 \\
(H_{CK} - H_{AK})/U_P &= -MRS_P P_K - P_{AK} > 0 \qquad (7.14) \\
(H_{CC} - H_{AC})/U_P &= -MRS_C > 0.
\end{aligned}
$$

Using these results we derive

$$
\begin{aligned}
\frac{\partial A}{\partial C} = \left.\frac{\dot{A}}{\dot{C}}\right|_{\dot{K}=0} &= \frac{H_{CC} - H_{AC}}{H_{AA} - H_{AC}} = \frac{-MRS_C}{MRS_P P_A + P_{AA}} > 0 \\
\frac{\partial A}{\partial K} = \left.\frac{\dot{A}}{\dot{K}}\right|_{\dot{C}=0} &= \frac{H_{CK} - H_{AK}}{H_{AA} - H_{AC}} = \frac{-MRS_P P_K - P_{AK}}{MRS_P P_A + P_{AA}} > 0.
\end{aligned} \qquad (7.15)
$$

So abatement expenditures increase with consumption and capital at an interior solution.

Turning to the $P = 0$-line we find from $dP = P_K dK + P_A dA = P_K dK + P_A \partial A/\partial C dC + P_A \partial A/\partial K dK$ that

$$
dP = \frac{-MRS_C P_A dC + (P_{AA} P_K - P_{AK} P_A) dK}{MRS_P P_A + P_{AA}}
$$

and thus for $dP = 0$

$$
\left.\frac{dC}{dK}\right|_{P=constant} = \frac{P_{AA} P_K - P_{AK} P_A}{MRS_C P_A} \qquad (7.16)
$$

which is positive (negative) if (7.1) is satisfied (violated). Above the $P = 0$-line there is more consumption and thus there would be more abatement at an interior solution (see (7.15)). Since negative pollution is not feasible, pollution is zero above the $P = 0$-line.

At every point of the $\tilde{C}$-line abatement expenditures are zero and pollution is positive. More abatement is necessary for zero

$P_{AA} = (-U_{CP} P_A + U_{PP} P_A^2 + U_P P_{AA})/U_P = (H_{AA} - H_{AC})/U_P$. Second, we derive $-MRS_P P_K - P_{AK} = (U_{CP} U_P P_K - U_{PP} U_C P_K)/U_P^2 - P_{AK} = (U_{CP} P_K - U_{PP} P_A P_K - U_P P_{AK})/U_P = (H_{CK} - H_{AK})/U_P$. Third, inserting (4.1) we obtain $-MRS_C = (U_{CC} U_P - U_{CP} U_C)/U_P^2 = (U_{CC} - U_{CP} P_A)/U_P = (H_{CC} - H_{AC})/U_P$.

pollution. Holding capital constant more consumption is therefore necessary (see (7.15)). Hence, the $P = 0$-line lies above the $\tilde{C}$-line except for the point of the $\tilde{C}$-line at $K = 0$ which lies on the $P = 0$-line because there $P(K, A) = P(0, 0) = 0$ (see figures 7.3 and 7.4). □

Proof of lemma 7.2

1. It follows from $\dot{K} = \phi(K) - \delta K - C - A(C, K) = 0$ that

$$\left.\frac{dC}{dK}\right|_{\dot{K}=0} = -\frac{\partial \dot{K}/\partial K}{\partial \dot{K}/\partial C} = \frac{\phi_K - \delta - \partial A/\partial K}{1 + \partial A/\partial C}. \tag{7.17}$$

Without pollution or if $A = 0$ we have $\partial A/\partial K = \partial A/\partial C = 0$. Thus below the $\tilde{C}$-line the $\dot{K} = 0$- and the $\dot{\bar{K}} = 0$-line are identical. The $\dot{\bar{K}} = 0$-line is rising (falling) if $K < K^*$ ($K > K^*$) where K^* is defined by $\phi_K(K^*) = \delta$. At an interior solution $\partial A/\partial K$ and $\partial A/\partial C$ are given in (7.15). For $P = 0$ we derive from $0 = dP = P_K dK + P_A dA$ that

$$\partial A/\partial K = -P_K/P_A > 0 \qquad \text{and} \qquad \partial A/\partial C = 0, \tag{7.18}$$

i.e. that we must follow an iso-pollution line in figure 7.1 whose slope is $-P_K/P_A$. Hence, the $\dot{K} = 0$-line is rising (falling) if $K < K'$ ($K > K'$) where K' is defined by $\phi_K(K') = \delta + \partial A/\partial K(C, K')$ where $\partial A/\partial K > 0$ is given in (7.15) or (7.18) and where K' depends on C. It follows that $K' < K^*$: The maximum of the $\dot{K} = 0$-line lies to the left of the maximum of the $\dot{\bar{K}} = 0$-line. When the $\dot{K} = 0$-line crosses the $\tilde{C}$-line from below, $\partial A/\partial K$ and $\partial A/\partial C$ suddenly become positive. Thus the $\dot{K} = 0$-line suddenly becomes flatter.[9] Above the $P = 0$-line an interior solution would

[9] *Selden* and *Song* (1995) disregard the kink of the $\dot{K} = 0$-line (and also of the $\dot{C} = 0$-line) using the $\dot{K} = 0$-line derived by *Forster* (1973) who, however, assumes away border solutions with zero abatement by assuming that $\lim_{A \to 0} |P_A| = \infty$. The removal of this assumption is the only difference between the model of *Selden* and *Song* (1995) and the model of *Forster* (1973).

prescribe negative pollution. As this is not feasible, we need less abatement to have zero pollution. Since the vertical distance between the $\dot{\bar{K}} = 0$- and the $\dot{K} = 0$-line are the abatement expenditures, it follows that this distance becomes smaller above the $P = 0$-line: The $\dot{K} = 0$-line becomes steeper when it crosses the $P = 0$-line from below (see figure 7.3). Below (above) the $\dot{K} = 0$-line consumption and thus abatement are smaller (greater) than on the line. So $\dot{K} = \phi(K) - \delta K - C - A$ is positive (negative) as indicated by the arrows in figures 7.2–7.4.

2. For $U_{CP} = 0$ we have $H_{AC} = H_{CK} = 0$ (see (7.4)). So inserting (7.4), (7.8), and (7.6) into the left hand side of (7.12) yields

$$\dot{C}|_{U_{CP}=0} = \frac{U_C}{U_{CC}} \left(\rho + \delta - \phi_K - \frac{P_K}{P_A + \gamma/(U_P + \mu)} \right). \tag{7.19}$$

The assumption $U_{CP} = 0$ has been used to derive (7.19) but will no longer be used below. (7.19) is zero if $M|_{U_{CP}=0} := \rho + \delta - \phi_K - P_K/\{P_A + \gamma/(U_P + \mu)\} = 0$. This can also be written as $\rho + \delta - \phi_K + P_K/MRS$ if $\mu = 0$ (see (7.10)). The last term in $M|_{U_{CP}=0}$, P_K/MRS, is zero for $C = 0$ since $\lim_{C \to 0} MRS = \infty$. Thus the $\dot{C}|_{U_{CP}=0} = 0$-line starts at point $(\hat{K}, 0)$ where $\phi_K(\hat{K}) = \rho + \delta$.

For $A = 0$ (and thus $\mu = 0$) we find

$$\left. \frac{\partial M|_{U_{CP}=0}}{\partial K} \right|_{A=0} = -\phi_{KK} + \frac{P_{KK} MRS - MRS_P P_K^2}{MRS^2} > 0 \tag{7.20}$$

$$\left. \frac{\partial M|_{U_{CP}=0}}{\partial C} \right|_{A=0} = \frac{-MRS_C P_K}{MRS^2} > 0. \tag{7.21}$$

For $A > 0$, $\gamma = 0$, and for any μ we obtain

$$\left. \frac{\partial M|_{U_{CP}=0}}{\partial K} \right|_{\gamma=0} = -\phi_{KK} \tag{7.22}$$

$$+ \frac{-P_{KK} P_A + P_{AK} P_K + (-P_{AK} P_A + P_{AA} P_K) \partial A/\partial K}{P_A^2}$$

$$\left.\frac{\partial M|_{U_{CP}=0}}{\partial C}\right|_{\gamma=0} = \frac{(-P_{AK}P_A + P_{AA}P_K)\partial A/\partial C}{P_A^2}. \tag{7.23}$$

At an interior solution (where $\gamma = \mu = 0$) we can insert (7.15) into (7.22) to find $-\phi_{KK} + \{-MRS_P P_{KK} P_A^2 - P_{AA}P_{KK}P_A + 2MRS_P P_{AK}P_A P_K + P_{AA}P_{AK}P_K + P_{AK}^2 P_A - MRS_P P_{AA}P_K^2 - P_{AA}P_{AK}P_K\}/\{P_A^2(MRS_P P_A + P_{AA})\}$ and thus

$$\left.\frac{\partial M|_{U_{CP}=0}}{\partial K}\right|_{\gamma=\mu=0} = -\phi_{KK} \tag{7.24}$$

$$+ \frac{MRS_P\left(-P_{KK} + 2P_{AK}\frac{P_K}{P_A} - P_{AA}\frac{P_K^2}{P_A^2}\right) - \frac{P_{AA}P_{KK}-P_{AK}^2}{P_A}}{MRS_P P_A + P_{AA}}$$

which is positive due to (7.2) and (7.3). For $P = 0$ we insert (7.18) into (7.23) and (7.22) to obtain $\frac{\partial M|_{P=0}}{\partial C} = 0$ and

$$\frac{\partial M|_{P=0}}{\partial K} = -\phi_{KK} + \frac{P_{KK} - 2P_{AK}P_K/P_A + P_{AA}P_K^2/P_A^2}{-P_A} > 0. \tag{7.25}$$

The slope of the $\dot{C}|_{U_{CP}=0} = 0$-line is given by $\left.\frac{dC}{dK}\right|_{\dot{C}|_{U_{CP}=0}=0} = -\left.\frac{\partial M/\partial K}{\partial M/\partial C}\right|_{U_{CP}=0}$. Therefore below the $\tilde{C}$-line the $\dot{C}|_{U_{CP}=0} = 0$-line is downward sloping (see (7.20) and (7.21)). At an interior solution the $\dot{C}|_{U_{CP}=0} = 0$-line is downward (upward) sloping if (7.1) is satisfied (violated, see (7.23) and (7.24)). Finally, above the $P = 0$-line the $\dot{C}|_{U_{CP}=0} = 0$-line is vertical (see (7.25)).

The $\dot{C}|_{U_{CP}=0} = 0$-line is continuous since it is always defined by $M|_{U_{CP}=0} = 0$, but it is kinked when it crosses the $\tilde{C}$- and the $P = 0$-line since its slope changes there (see figures 7.2–7.4). Even if the $\dot{C}|_{U_{PC}=0} = 0$-line is upward sloping at an interior solution, it always lies to the left of the $\dot{\tilde{C}} = 0$-line: On the $\dot{\tilde{C}} = 0$-line (where $\phi_K = \rho + \delta$ holds) $M|_{U_{PC}=0} = \rho + \delta - \phi_K - P_K/P_A = -P_K/P_A$ is always positive. Thus the $\dot{C}|_{U_{PC}=0} = 0$-line never crosses the $\dot{\tilde{C}} = 0$-line.

3. To derive the $\dot{C} = 0$-line above the $P = 0$-line we insert (7.18) and (7.4) into the left hand side of (7.12) to obtain $H_{CC}\dot{C} - H_{AC}P_K/P_A\dot{K} + H_{CK}\dot{K} = H_{CC}\dot{C} = \dot{\lambda}$. Thus (7.19) ensues even for $U_{CP} \neq 0$: Above the $P = 0$-line the $\dot{C} = 0$- and the $\dot{C}|_{U_{PC}=0} = 0$-line are identical. Since $M|_{P=0}$ increases with capital (see (7.25)), it follows from (7.19) and $U_{CC} < 0$ (see (7.5)) that consumption rises (falls) to the left (right) of the $\dot{C} = 0$-line.

Below the $\tilde{C}$-line $A = \dot{A} = 0$. Thus for $U_{CP} \neq 0$ we insert (7.4), (7.8), and (7.6) into the left hand side of (7.12) to find

$$\dot{C}|_{A=0} = \frac{U_C}{U_{CC}}\left(\rho + \delta - \phi_K + \frac{P_K}{MRS}\right) - \frac{U_{CP}P_K}{U_{CC}}\dot{K} \tag{7.26}$$

which is zero if $M|_{A=0} := U_C(\rho+\delta-\phi_K+P_K/MRS) - U_{CP}P_K\dot{K} = 0$. For an interior solution we insert (7.13) into the left hand side of (7.12) to obtain $\{\dot{C}[H_{CC}(H_{AA} - H_{AC}) + H_{AC}(H_{CC} - H_{AC})] + \dot{K}[H_{CK}(H_{AA} - H_{AC}) + H_{AC}(H_{CK} - H_{AK})]\}/\{H_{AA} - H_{AC}\} = \dot{\lambda}$. Together with (7.8) and (7.6) this yields

$$\begin{aligned}\dot{C} &= \tfrac{U_C(H_{AA}-H_{AC})}{H_{AA}H_{CC}-H_{AC}^2}\left(\rho + \delta - \phi_K - \tfrac{P_K}{P_A}\right) + \tfrac{H_{AC}H_{AK}-H_{AA}H_{CK}}{H_{AA}H_{CC}-H_{AC}^2}\dot{K} \\ &= \tfrac{U_CU_P(MRS_PP_A+P_{AA})}{H_{AA}H_{CC}-H_{AC}^2}\left(\rho + \delta - \phi_K - \tfrac{P_K}{P_A}\right) \\ &\quad + \tfrac{U_{CP}U_P(P_{AK}P_A-P_{AA}P_K)}{H_{AA}H_{CC}-H_{AC}^2}\dot{K}\end{aligned} \tag{7.27}$$

where we have inserted (7.14) and (7.4). It immediately follows that $\dot{C} = 0$, if $M := U_C(MRS_PP_A + P_{AA})(\rho+\delta-\phi_K - P_K/P_A) + U_{CP}(P_{AK}P_A - P_{AA}P_K)\dot{K} = 0$. We can easily pin down the $\dot{C} = 0$-line at two points. On the one hand, the first term in $M|_{A=0}$ can also be written as $U_C(\rho+\delta-\phi_K) - U_PP_K$. Since $U_C = \infty$ for $C = 0$ (as $\lim_{C\to 0} MRS = \infty$), all other terms in $M|_{A=0}$ are dwarfed by $U_C(\rho+\delta-\phi_K)$. This term is only zero at $K = \hat{K}$. Thus the $\dot{C} = 0$-line starts at $(\hat{K}, 0)$, i.e. at the same point as the $\dot{C}|_{U_{CP}=0} = 0$-line. On the other hand, both $M|_{A=0}$ and M are zero for the same values as before if $\dot{K} = 0$. Hence, the $\dot{C} = 0$-line intersects the $\dot{K} = 0$-line at the intersection point of the $\dot{C}|_{U_{CP}=0} = 0$-line with the $\dot{K} = 0$-line and only there. So the steady state is determined by the $\dot{C}|_{U_{CP}=0} = 0$- and the $\dot{K} = 0$-line.

The derivatives of $M|_{A=0}$ and M with respect to C and K involve third derivatives whose signs are unknown. If all four derivatives are positive, the $\dot{C} = 0$-line is falling. However, if some derivative of $M|_{A=0}$ or M with respect to C or K becomes negative, the $\dot{C} = 0$-line has rising parts or might even be a wavy line. In figures 7.2–7.4 we assume that the $\dot{C} = 0$-line is not a wavy line.

The $\dot{C} = 0$-line divides the phase diagram in two unequal halves (see figures 7.2–7.4). Let us analyse whether consumption is rising or falling in the left half. Consider two points – points X and Y in figure 7.2 – on the $\dot{K} = 0$-line at the same consumption level above (or below) the $\tilde{C}$-line. It follows from (7.24) (or (7.20)) that $M|_{U_{CP}=0} = \rho + \delta - \phi_K + P_K/MRS$ is negative at point X, but positive at point Y. Thus we derive from (7.27) (or (7.26)) that at point X (where $\dot{K} = 0$) $\dot{C}$ (or $\dot{C}|_{A=0}$) is positive, but at point Y it is negative. Hence, consumption rises in the left half, but falls in the right half.

Therefore whenever $\dot{C}$ or $\dot{C}|_{A=0}$ is negative at a point of the $\dot{C}|_{U_{CP}=0} = 0$-line, the $\dot{C} = 0$-line lies to the left of the $\dot{C}|_{U_{CP}=0} =$ 0-line. At a point on the $\dot{C}|_{U_{CP}=0} = 0$-line $\dot{C}$ or $\dot{C}|_{A=0}$ simplifies to $U_{CP}U_P(P_{AK}P_A - P_{AA}P_K)\dot{K}/(H_{AA}H_{CC} - H_{AC}^2)$ or $\frac{-U_{CP}P_K}{U_{CC}}\dot{K}$ where $H_{AA}H_{CC} - H_{AC}^2 > 0$ and $U_{CC} < 0$ (see (7.5)). So below the $\tilde{C}$-line the $\dot{C}|_{A=0} = 0$-line lies to the left of the $\dot{C}|_{U_{CP}=0} = 0$-line, if $\dot{C}|_{A=0}$ is negative, i.e. if $U_{CP} < 0$ and $\dot{K} > 0$ (as below the $\dot{K} = 0$-line). At an interior solution the $\dot{C} = 0$-line lies to the left of the $\dot{C}|_{U_{CP}=0} = 0$-line, if $\dot{C}$ is negative, i.e. if $U_{CP}(P_{AK}P_A - P_{AA}P_K)\dot{K} > 0$, for example if $U_{CP} < 0$, (7.1) is satisfied, and $\dot{K} > 0$. If the $\dot{C} = 0$-line lies to the left of the $\dot{C}|_{U_{CP}=0} = 0$-line below the $\dot{K} = 0$-line, it lies to the right of the $\dot{C}|_{U_{CP}=0} = 0$-line above the $\dot{K} = 0$-line as shown in figures 7.2 and 7.4.

It remains to be proven that the $\dot{C} = 0$-line is discontinuous when it crosses the $\tilde{C}$- and the $P = 0$-line. At the point on the $\tilde{C}$-line where $M|_{A=0} = 0$ holds we have $U_C(\rho + \delta - \phi_K - P_K/P_A) = U_{CP}P_K\dot{K}$. Thus $\dot{C}$ simplifies to $U_{CP}P_K\dot{K}(MRS_P P_A + P_{AK}P_A/P_K)U_P/(H_{AA}H_{CC} - H_{AC}^2)$ or U_{CP} multiplied by a negative term if $\dot{K} > 0$. For $U_{CP} < 0$ ($U_{CP} > 0$) we find $\dot{C} > 0$ ($\dot{C} < 0$) so that this point of the $\dot{C}|_{A=0} = 0$-line lies to the left (right) of the $\dot{C} = 0$-line as shown in figures 7.2 and 7.4 (7.3). Finally, since for $U_{CP} \neq 0$ the $\dot{C} = 0$-line lies to the left or to the right of the $\dot{C}|_{U_{CP}=0} = 0$-line, this is also the case at the intersection of the $\dot{C}|_{U_{CP}=0} = 0$-line with the $P = 0$-line as shown in figures 7.3 and 7.4. □

Proof of lemma 7.3

1. We first analyse an interior steady state and assume that (7.1) holds. The local behaviour at the steady state (SS) is determined by the Jacobian matrix J. Since at each steady state $\dot{K} = \dot{C} = 0$ holds, (7.27) shows that $\rho + \delta - \phi_K - P_K/P_A = 0$. Thus – using that $\partial A/\partial K$, $\partial A/\partial C$, and the derivatives of $M|_{U_{CP}=0}$ are all positive and given in (7.15), (7.23), and (7.24) – it follows from $\dot{K} = \phi(K) - \delta K - C - A(C, K)$ and (7.27) that

$$
\begin{aligned}
\left.\frac{\partial \dot{K}}{\partial K}\right|_{SS} &= \phi_K - \delta - \frac{\partial A}{\partial K} = \rho - \frac{P_K}{P_A} + \frac{MRS_P P_K + P_{AK}}{MRS_P P_A + P_{AA}} \\
&= \rho + \frac{-P_K P_{AA} + P_A P_{AK}}{P_A(MRS_P P_A + P_{AA})} > 0 \\
\left.\frac{\partial \dot{K}}{\partial C}\right|_{SS} &= -1 - \frac{\partial A}{\partial C} < 0 \\
\left.\frac{\partial \dot{C}}{\partial K}\right|_{SS} &= \frac{U_C U_P (MRS_P P_A + P_{AA})}{H_{AA}H_{CC} - H_{AC}^2} \left.\frac{\partial M|_{U_{CP}=0}}{\partial K}\right|_{\gamma=\mu=0} \\
&\quad + \frac{U_{CP}U_P(P_{AK}P_A - P_{AA}P_K)}{H_{AA}H_{CC} - H_{AC}^2} \left.\frac{\partial \dot{K}}{\partial K}\right|_{SS} \\
\left.\frac{\partial \dot{C}}{\partial C}\right|_{SS} &= \frac{U_C U_P (MRS_P P_A + P_{AA})}{H_{AA}H_{CC} - H_{AC}^2} \left.\frac{\partial M|_{U_{CP}=0}}{\partial C}\right|_{\gamma=0} \\
&\quad + \frac{U_{CP}U_P(P_{AK}P_A - P_{AA}P_K)}{H_{AA}H_{CC} - H_{AC}^2} \left.\frac{\partial \dot{K}}{\partial C}\right|_{SS}.
\end{aligned}
$$

The determinant of the Jacobian $J_{SS} = \begin{bmatrix} \frac{\partial \dot{K}}{\partial K}\Big|_{SS} & \frac{\partial \dot{K}}{\partial C}\Big|_{SS} \\ \frac{\partial \dot{C}}{\partial K}\Big|_{SS} & \frac{\partial \dot{C}}{\partial C}\Big|_{SS} \end{bmatrix}$ simplifies to

$$\begin{aligned} \det J_{SS} &= \frac{U_C U_P (MRS_P P_A + P_{AA})}{H_{AA} H_{CC} - H_{AC}^2} \left[\left(\rho + \frac{-P_K P_{AA} + P_A P_{AK}}{P_A (MRS_P P_A + P_{AA})} \right) \right. \\ &\quad \left. \cdot \frac{\partial M|_{U_{CP}=0}}{\partial C}\Big|_{\gamma=0} + \left(1 + \frac{\partial A}{\partial C}\right) \frac{\partial M|_{U_{CP}=0}}{\partial K}\Big|_{\gamma=\mu=0} \right]. \end{aligned} \quad (7.28)$$

Thus it follows from (7.5) that $\det J_{SS} < 0$. Therefore the steady state is a saddle point.

If (7.1) is violated, we can see in figure 7.4 that the $\dot{C}|_{U_{CP}=0} = 0$-line is always more positively sloped than the $\dot{K} = 0$-line at the steady state, i.e. $-\frac{\partial M/\partial K}{\partial M/\partial C}\Big|_{U_{CP}=0} > (\phi_K - \delta - \partial A/\partial K)/(1 + \partial A/\partial C)$ (see (7.23), (7.24), and (7.17)). Using $\frac{\partial \dot{K}}{\partial K}\Big|_{SS}$ from above, multiplying through by $(1 + \partial A/\partial C)\partial M|_{U_{CP}=0}/\partial C < 0$, and rearranging we find that

$$\left(\rho + \frac{-P_K P_{AA} + P_A P_{AK}}{P_A (MRS_P P_A + P_{AA})} \right) \frac{\partial M|_{U_{CP}=0}}{\partial C}\Big|_{\gamma=0} + \left(1 + \frac{\partial A}{\partial C}\right) \frac{\partial M|_{U_{CP}=0}}{\partial K}\Big|_{\gamma=\mu=0}$$

is positive. It follows from (7.28) that $\det J_{SS} < 0$. The steady state is also a saddle point.

If the steady state lies below the $\tilde{C}$-line, we use $\partial A/\partial K = \partial A/\partial C = 0$ and (7.26) instead of (7.27). The determinant looks similar

$$\det J_{SS}^{A=0} = \frac{U_C}{U_{CC}} \left[\left(\rho + \frac{P_K}{MRS} \right) \frac{\partial M|_{U_{CP}=0}}{\partial C}\Big|_{A=0} + \frac{\partial M|_{U_{CP}=0}}{\partial K}\Big|_{A=0} \right] < 0$$

where the sign follows from (7.5), (7.20), and (7.21). Thus the zero-abatement steady state is also a saddle point.

For a zero pollution steady state we use (7.19) which is also valid for $P = 0$ and $\partial A/\partial K = -P_K/P_A$ and $\partial A/\partial C = 0$ (see (7.18)) to obtain $\frac{\partial \dot{K}}{\partial K}\Big|_{SS} = \phi_K - \delta - \partial A/\partial K = \rho$, $\frac{\partial \dot{K}}{\partial C}\Big|_{SS} = -1$, $\frac{\partial \dot{C}}{\partial K}\Big|_{SS} = U_C/U_{CC}\ \partial M|_{P=0}/\partial K$ where $\partial M|_{P=0}/\partial K > 0$ is derived in (7.25), and $\left.\frac{\partial \dot{C}}{\partial C}\right|_{SS} = 0$. Thus we find $\det J_{SS}^{P=0} = U_C/U_{CC} \cdot \partial M|_{P=0}/\partial K < 0$ since $U_{CC} < 0$. So the zero pollution steady state is also a saddle point.

2. Along an optimal path the costate variable λ must be a continuous function of time (*Feichtinger* and *Hartl* 1986, 40). Thus sudden changes of the control variable, say from C_1 to C_2, can only be optimal, if λ takes the same value at C_1 and C_2. According to (7.6) $\lambda = U_C(C, P(K, A))$. At a corner solution with $A = 0$ or $P = 0$ we find $\partial\lambda/\partial C = U_{CC} < 0$. At an interior solution we use (7.4) and (7.15) to derive $\partial\lambda/\partial C = U_{CC} + U_{CP}P_A\partial A/\partial C = H_{CC} + H_{AC}\partial A/\partial C = (H_{AA}H_{CC} - H_{AC}^2)/(H_{AA} - H_{AC}) < 0$ where the sign follows from (7.5) and (7.14). So λ falls with rising consumption and sudden changes of consumption are never optimal. Note that if consumption never changes discontinuously, abatement does neither since they are tied together by (7.10).

3. The proof that the transversality condition holds on the saddle path is analogous to the proof of lemma 6.2.3: It is clear from its derivation that the saddle path satisfies the first order conditions (7.6) – (7.8). To show that the transversality condition (7.9) also holds, we use (7.8) to write $\lambda(t)$ as

$$\begin{aligned}\lambda(t) &= \lambda(0)\exp\left\{\int_0^t \rho + \delta - \phi_K(\check{K}(t))\right. \\ &\qquad \left. - \frac{P_K(\check{K}(t), \check{A}(t))}{P_A(\check{K}(t), \check{A}(t)) + \frac{\check{\gamma}(t)}{U_P[\check{C}(t), P(\check{K}(t), \check{A}(t))] + \check{\mu}(t)}} dt\right\}\end{aligned}$$

where a check ($\check{\cdot}$) means the variable on the saddle path. Hence, the transversality condition (7.9) becomes

$$\begin{aligned}\lim_{t\to\infty} \lambda(0)\ [K(t) - \check{K}(t)]\exp&\left\{\int_0^t \delta - \phi_K(\check{K}(t))\right. \\ &\left. - \frac{P_K(\check{K}(t), \check{A}(t))}{P_A(\check{K}(t), \check{A}(t)) + \frac{\check{\gamma}(t)}{U_P[\check{C}(t), \check{P}(t)] + \check{\mu}(t)}} dt\right\} \geq 0.\end{aligned}$$

The saddle path lies always to the left of the $\dot{C}|_{U_{CP}=0} = 0$-line on which $\rho + \delta - \phi_K - P_K/[P_A + \gamma/(U_P + \mu)] = 0$ (see figure 7.2–7.4). So $\delta - \phi_K - P_K/[P_A + \gamma/(U_P + \mu)] \leq -\rho$ holds along the whole saddle path (see (7.20), (7.24), and (7.25)). Hence, because $K - \check{K}$ is bounded, the limit is zero and the transversality condition is satisfied.

The proof that every other path violates either the first order conditions or the transversality condition is exactly analogous to the proof of lemma 6.2.3 (see page 158).

4. To prove that the saddle path gives maximal, not minimal utility, we must show that H^0 is concave in K at all points of time, where H^0 is the Hamiltonian with the optimal values of C and A inserted for a given K (*Feichtinger* and *Hartl* 1986, 43).

For $A = 0$ we have $H^0 = U[C(K), P(K, 0)] + \lambda[\phi(K) - \delta K - C(K)]$ and therefore $H^0_K = H_C \partial C/\partial K + H_K$ where $H_C = 0$ (see (7.6)). So the sufficient condition for a maximal path is

$$H^0_{KK} = H_{CC}\,(\partial C/\partial K)^2 + 2H_{CK}\partial C/\partial K + H_{KK} \leq 0.$$

Because D^2H (as defined on page 167) is negative semidefinite, $z'[D^2H]z \leq 0$ holds for all vectors z. Taking $z' = [\partial C/\partial K,\ \ 0,\ \ 1]$ shows that $H^0_{KK} \leq 0$ is satisfied.

At an interior solution H^0 is $U\{C(K), P[K, A(C(K), K)]\} + \lambda[\phi(K) - \delta K - C(K) - A(C(K), K)]$. Thus H^0_K equals $H_C \partial C/\partial K + H_A \partial A/\partial C\ \partial C/\partial K + H_A \partial A/\partial K + H_K$ where $H_C = H_A = 0$ according to (7.6) and (7.7). Therefore we must have

$$\begin{aligned} H^0_{KK} &= \left(\tfrac{\partial C}{\partial K}\right)^2 \left[H_{CC} + 2H_{AC}\tfrac{\partial A}{\partial C} + H_{AA}\left(\tfrac{\partial A}{\partial C}\right)^2\right] \\ &\quad + 2\tfrac{\partial C}{\partial K}\left[H_{CK} + H_{AK}\tfrac{\partial A}{\partial C} + H_{AC}\tfrac{\partial A}{\partial K} + H_{AA}\tfrac{\partial A}{\partial C}\tfrac{\partial A}{\partial K}\right] \\ &\quad + \left[H_{KK} + 2H_{AK}\tfrac{\partial A}{\partial K} + H_{AA}\left(\tfrac{\partial A}{\partial K}\right)^2\right] \leq 0. \end{aligned}$$

Taking $z' = [\partial C/\partial K,\ \ \partial A/\partial C\ \partial C/\partial K + \partial A/\partial K,\ \ 1]$ we find that $H^0_{KK} = z'[D^2H]z \leq 0$ is satisfied.

For zero pollution H^0 is given by $U[C(K),0]+\lambda[\phi(K)-\delta K - C(K)-A(K)]$. Thus we derive $H^0_K = (U_C[C(K),0]-\lambda)\partial C/\partial K + \lambda(\phi_K - \delta + P_K/P_A)$ where (7.18) has been inserted. Using (6.3) and $H_C = U_C - \lambda = 0$ (see (7.6)) the sufficient condition is found to be

$$H^0_{KK} = U_{CC}\left(\partial C/\partial K\right)^2 + U_C\Big[\phi_{KK} + \left(P_{KK} - 2P_{AK}P_K/P_A + P_{AA}P_K^2/P_A^2\right)/P_A\Big] \leq 0$$

which is satisfied due to (7.3) and (7.5).

Finally, H^0 is also concave when the optimal path crosses the $\tilde{C}$- and the $P=0$-line because below the $P=0$-line H^0_K is defined by $H_K = U_P P_K + \lambda(\phi_K - \delta)$ and above the $P=0$-line H^0_K is given by $\lambda(\phi_K - \delta + P_K/P_A)$ which on the $P=0$-line (where $\mu = \gamma = 0$) equals $\lambda(\phi_K - \delta + U_P P_K/U_C) = \lambda(\phi_K - \delta) + U_P P_K$ (see (7.10) and (7.6)). Thus the slope of H^0 does not change when the optimal path crosses the $\tilde{C}$- or the $P=0$-line. Only the curvature changes there, but remains concave. □

Proof of lemma 7.4

It can easily be seen that the constant-pollution lines (including the $P=0$-line), the $\tilde{C}$-, the $\dot{K}=0$-, the $\dot{C}|_{U_{CP}=0}=0$-line, and thus the steady state are all robust to monotonic transformations of the felicity function (see (7.16), (7.11), (7.17) combined with (7.15) or (7.18), and (7.19)). This holds also true for the $\dot{C}=0$-line above the $P=0$-line since it is identical to the $\dot{C}|_{U_{CP}=0}=0$-line above the $P=0$-line (see lemma 7.2.3). Below the $P=0$-line the $\dot{C}=0$-line is no longer robust because it depends on U_{CP}/U_C (see $M|_{A=0}$ and M defined after (7.26) and (7.27)): For instance, using the same monotonic transformation as in the proof of lemma 6.3, i.e. using $\tilde{U} = U^x$ where $x \neq 1$ and $x, U > 0$, we find that U_{CP}/U_C is not robust as $\tilde{U}_{CP}/\tilde{U}_C = U_{CP}/U_C + (x-1)U_P/U \neq U_{CP}/U_C$ (see (6.18)). Furthermore, whereas the level of $\dot{K} = \phi(K) - \delta K - C - A$

at some point in the phase diagram is robust to changes in the elasticity of intertemporal substitution, we will show in the proof of proposition 7.1.2 below that the level of $\dot{C}$ is not robust. Thus if $\dot{C}$ is relatively low (high) everywhere, we must start with a high (low) consumption level at K_0 to reach the steady state as shown by path c (a) in figure 7.6. □

Proof of proposition 7.1.2

Most of the proof is given in the main text. Here we only study an example of a monotonic transformation of the felicity function for which the elasticity of intertemporal substitution falls. For this example we show, first, that (7.5) remains valid, second, that D^2H as defined on page 167 remains negative semidefinite, and, third, that the optimal path shifts upwards.

Consider the same monotonic transformation of the felicity function as in the proof of lemma 6.3, i.e. consider $\tilde{U} = U^x$ where $0 < x < 1$ and $U > 0$ (if $U < 0$, we use $a > 0$ such that $U + a > 0$ and define $\tilde{U} = (U + a)^x$). As in chapter 6 the whole proof remains valid if we instead consider $\tilde{U} = -U^x$ where $x < 0$. Since this transformation makes the differences between felicity levels attained at different points in time smaller, the elasticity of intertemporal substitution falls and individuals prefer a smooth development of felicity levels along the optimal path.

To show that the transformed felicity function $\tilde{U}$ fulfils (7.5), we must ensure that $\tilde{H}_{CC} = \tilde{U}_{CC} < 0$ and $\tilde{H}_{AA}\tilde{H}_{CC} - \tilde{H}^2_{AC} > 0$ both hold (if $U_{CC} < 0$ and $H_{AA}H_{CC} - H^2_{AC} > 0$). As in the proof of lemma 6.3 we derive from $\tilde{U} = U^x$ that

$$\begin{aligned}
\tilde{U}_C &= xU^{x-1}U_C > 0 \\
\tilde{U}_P &= xU^{x-1}U_P > 0 \\
\tilde{H}_{CC} = \tilde{U}_{CC} &= xU^{x-1}U_{CC} + x(x-1)U^{x-2}U_C^2 < 0 \\
\tilde{U}_{PP} &= xU^{x-1}U_{PP} + x(x-1)U^{x-2}U_P^2 \\
\tilde{U}_{CP} &= xU^{x-1}U_{CP} + x(x-1)U^{x-2}U_CU_P.
\end{aligned}$$

Using (7.4) and defining $\bar{H}_{KK} := U_{PP}P_K^2 + U_P P_{KK} = H_{KK} - \lambda\phi_{KK}$ we obtain

$$\begin{array}{rcl}
\tilde{H}_{AA} & = & xU^{x-1}H_{AA} + x(x-1)U^{x-2}U_P^2P_A^2 \\
\tilde{H}_{KK} & = & xU^{x-1}\bar{H}_{KK} + x(x-1)U^{x-2}U_P^2P_K^2 + \lambda\phi_{KK} \\
\tilde{H}_{AC} & = & xU^{x-1}H_{AC} + x(x-1)U^{x-2}U_CU_PP_A \\
\tilde{H}_{AK} & = & xU^{x-1}H_{AK} + x(x-1)U^{x-2}U_P^2P_AP_K \\
\tilde{H}_{CK} & = & xU^{x-1}H_{CK} + x(x-1)U^{x-2}U_CU_PP_K.
\end{array} \tag{7.29}$$

For the second equality sign of the next derivation we use $U_C = U_P P_A$ which holds at an interior solution (see (7.10)). We are allowed to use this equation because $H_{AA}H_{CC} - H_{AC}^2 > 0$ is only a necessary condition for the interior solution as the previous proofs have shown. Therefore we find

$$\begin{array}{rl}
\multicolumn{2}{l}{(\tilde{H}_{AA}\tilde{H}_{CC} - \tilde{H}_{AC}^2)/(xU^{x-1})^2} \\
= & (H_{AA} + U_P^2P_A^2(x-1)/U)(H_{CC} + U_C^2(x-1)/U) \\
 & -(H_{AC} + U_CU_PP_A(x-1)/U)^2 \\
= & (H_{AA} + U_CU_PP_A(x-1)/U)(H_{CC} + U_CU_PP_A(x-1)/U) \\
 & -(H_{AC} + U_CU_PP_A(x-1)/U)^2 \\
= & H_{AA}H_{CC} - H_{AC}^2 \\
 & +U_CU_PP_A(H_{AA} - H_{AC} + H_{CC} - H_{AC})(x-1)/U > 0
\end{array} \tag{7.30}$$

where the sign follows from (7.14). Hence, (7.5) is also satisfied for $\tilde{U}$.

Next, we must prove that $D^2\tilde{H}$ is negative semidefinite assuming that the original D^2H is negative semidefinite – or, more precisely, assuming that $D^2\bar{H}$ as defined in the next equation is negative semidefinite. Using (7.29) we can write $D^2\tilde{H}$ as

$$D^2\tilde{H} = xU^{x-1}\underbrace{\begin{bmatrix} H_{CC} & H_{AC} & H_{CK} \\ H_{AC} & H_{AA} & H_{AK} \\ H_{CK} & H_{AK} & \bar{H}_{KK} \end{bmatrix}}_{:=D^2\bar{H}}$$

$$
+x(x-1)U^{x-2}\underbrace{\begin{bmatrix} U_C^2 & U_C U_P P_A & U_C U_P P_K \\ U_C U_P P_A & U_P^2 P_A^2 & U_P^2 P_A P_K \\ U_C U_P P_K & U_P^2 P_A P_K & U_P^2 P_K^2 \end{bmatrix}}_{:=W} + \begin{bmatrix} 0 & 0 & 0 \\ 0 & 0 & 0 \\ 0 & 0 & \lambda\phi_{KK} \end{bmatrix}.
$$

To show that $D^2\tilde{H}$ is negative semidefinite we must make sure that $z'[D^2\tilde{H}]z \leq 0$ for all vectors $z' = [z_1, \quad z_2, \quad z_3]$. Hence, we must have

$$
z'[D^2\tilde{H}]z = xU^{x-1}z'[D^2\bar{H}]z + x(x-1)U^{x-2}z'Wz + z_3^2\lambda\phi_{KK} \leq 0.
$$

Since we know that $z'[D^2\bar{H}]z \leq 0$ and $\lambda\phi_{KK} < 0$, this is satisfied if we can show that $z'Wz \geq 0$. Since W can be written as $W = vv'$ where $v' = [U_C, \quad U_P P_A, \quad U_P P_K]$, we find

$$
z'Wz = z'[vv']z = (z'v)(v'z) = (z_1 U_C + z_2 U_P P_A + z_3 U_P P_K)^2 \geq 0.
$$

This completes the proof that $D^2\tilde{H}$ is negative semidefinite.

It remains to be proven that the optimal path moves upwards when we apply this monotonic transformation of the felicity function. To do so we consider the equations for $\dot{C}$: If we can show that $\dot{C}$ becomes smaller in the relevant area (i.e. below the $\dot{K} = 0$-line and to the left of the $\dot{C} = 0$-line), then we must start with a higher consumption level at K_0 to reach the steady state. Hence, the optimal path shifts upwards. We first analyse $\dot{C}$ above the $\dot{P} = 0$-line which is given in (7.19). The term $|U_C/U_{CC}|$ in (7.19) becomes smaller because

$$
\left|\frac{\tilde{U}_C}{\tilde{U}_{CC}}\right| = \left|\frac{U_C}{U_{CC} + U_C^2(x-1)/U}\right| < \left|\frac{U_C}{U_{CC}}\right|.
$$

Hence, $\dot{C}$ becomes smaller in the relevant area above the $\dot{P} = 0$-line.

Next, consider the corner solution with zero abatement expenditures. $\dot{C}|_{A=0}$ is given in (7.26) and consists of two terms.

The first is identical to the term analysed above and therefore decreases as well. The second term becomes more negative (in the relevant area where $\dot{K} > 0$) as $-\tilde{U}_{CP}/\tilde{U}_{CC} < -U_{CP}/U_{CC}$ because

$$\begin{aligned}
\frac{\tilde{U}_{CP}}{\tilde{U}_{CC}} = \frac{U_{CP} + U_C U_P (x-1)/U}{U_{CC} + U_C^2 (x-1)/U} &> \frac{U_{CP}}{U_{CC}} \\
U_{CC}U_{CP} + U_{CC}U_C U_P (x-1)/U &> U_{CC}U_{CP} + U_{CP}U_C^2 (x-1)/U \\
U_{CC}U_P - U_{CP}U_C &> 0
\end{aligned}$$

which holds according to (4.1). So in the relevant area below the $\tilde{C}$-line $\dot{C}|_{A=0}$ becomes also smaller.

Finally, we have to analyse $\dot{C}$ at an interior solution. Then $\dot{C}$ is given in (7.27) and consists again of two terms. The first term declines because $|U_C U_P/(H_{AA}H_{CC} - H_{AC}^2)|$ becomes smaller as

$$\begin{aligned}
\left|\frac{\tilde{U}_C \tilde{U}_P}{\tilde{H}_{AA}\tilde{H}_{CC} - \tilde{H}_{AC}^2}\right| &= \left|\frac{U_C U_P}{H_{AA}H_{CC} - H_{AC}^2 + U_C U_P P_A (H_{AA} + H_{CC} - 2H_{AC})(x-1)/U}\right| \\
&< \left|\frac{U_C U_P}{H_{AA}H_{CC} - H_{AC}^2}\right|
\end{aligned}$$

where we have used (7.30). The second term in (7.27) becomes more negative because a constant negative term is multiplied by a term which becomes more positive since

$$\begin{aligned}
&\frac{\tilde{U}_{CP}\tilde{U}_P}{\tilde{H}_{AA}\tilde{H}_{CC} - \tilde{H}_{AC}^2} \\
&= \frac{U_{CP}U_P + U_C U_P^2 (x-1)/U}{H_{AA}H_{CC} - H_{AC}^2 + U_C U_P P_A (H_{AA} + H_{CC} - 2H_{AC})(x-1)/U} > \frac{U_{CP}U_P}{H_{AA}H_{CC} - H_{AC}^2}
\end{aligned}$$

$$\begin{aligned}
U_C U_P^2 (H_{AA}H_{CC} - H_{AC}^2) &> U_C U_P^2 U_{CP} P_A (H_{AA} + H_{CC} - 2H_{AC}) \\
H_{CC}(H_{AA} - H_{AC}) + H_{AC}(H_{CC} - H_{AC}) &> H_{AC}(H_{AA} + H_{CC} - 2H_{AC}) \\
(H_{CC} - H_{AC})(H_{AA} - H_{AC}) &> 0
\end{aligned}$$

where the sign follows from (7.14) and where we have made use of $U_{CP}P_A = H_{AC}$ (see (7.4)). So at an interior solution $\dot{C}$ becomes also smaller in the relevant area. Therefore the optimal path shifts upwards when we transform the felicity function according to $\tilde{U} = U^x$. □

Part III

The Environmental Kuznets Curve and Flow versus Stock Pollution

8. The Neglect of Future Damages: Myopic Governments

8.1 Introduction

The main purpose of part III of this book is to find a possible explanation of the empirical evidence that the PIR is an EKC for flow pollutants, but that the PIR is monotonically rising for stock pollutants. As discussed in subsection 2.3.2 the EKC is found for sulphur dioxide (SO_2), suspended particulate matter (SPM), oxides of nitrogen (NO_x), carbon monoxide (CO), volatile organic compounds (VOC), and for some (but not all) sorts of river pollution (see tables 2.1–2.3). These are all flow pollutants. In contrast, for the stock pollutants carbon dioxide (CO_2) and municipal waste the estimated PIR is monotonically rising (see tables 2.2 and 2.4). We find a new explanation of the EKC in chapters 8 and 9: The EKC for a flow pollutant might be due to the neglect of future damages and due to ever rising stock pollution.

We develop an overlapping generations model based on the model of *John* and *Pecchenino* (1994). In this chapter we consider the solution of a succession of myopic governments and in chapter 9 we crudely sketch the far-sighted optimum. We extend the model of *John* and *Pecchenino* (1994) in several directions: First, since in the myopic solution every generation solves a static problem (where the solution determines the initial conditions of the next generation), we can apply our method developed in section 4.2, i.e. we make only ordinal, but no cardinal assumptions about the utility function. Second, we use a completely different pollution function than *John* and *Pecchenino* (1994). Third, we again restrict pollution to be nonnegative. Fourth, while there is only one pollutant in the model of *John* and *Pecchenino* (1994), we extend the model to two pollutants. To my knowledge this is the first model of the EKC with two pollutants.

In the model the economy is equipped with a production technology which causes emissions of a flow and of a stock pollutant. On the one hand, the flow pollutant has an immediate and only an immediate effect on the environment. On the other hand, the stock pollutant harms the environment only in the future since stock pollutants frequently need time to accumulate before the damage occurs. Furthermore, there are two abatement technologies – one for each pollutant. Myopic governments never abate the stock pollutant because the positive effects are only felt in the future which is of no concern for myopic governments. Therefore we show that on the path followed by myopic governments the PIR for the flow pollutant is an EKC while the PIR for the stock pollutant is monotonically rising. This is consistent with the empirical evidence. The EKC stems from the fact that myopic governments abate flow pollution in order to keep the aggregate damage of both pollutants at a reasonably low level. Because stock pollution increases, abatement of the flow pollutant rises so fast that flow pollution declines. If there is no stock pollutant, however, it is possible that flow pollution rises monotonically with income. We also find that the turning point of the EKC for the flow pollutant depends on the level of stock and flow pollution.

The model also takes into account that abatement of a flow pollutant often causes higher emissions of a stock pollutant or in other words that abatement of flow pollution increases stock pollution (cf. section 3.3). This is often observed in reality as the following two examples show: First, in the catalytic converter of a car the exhaust components – CO, NO_x, and Hydrocarbons which are all flow pollutants[1] – are all transformed into the stock pollutant CO_2

[1] As mentioned in subsection 2.3.2 the lifetime of CO is 1–3 months, that of NO_x is 2–5 days, and that of Hydrocarbons ranges from hours to a few days – except for methane which is a stock pollutant with a lifetime of 8–12 years (*Liu* and *Lipták* 2000, 32 and *IPCC* 1996, 92).

(*Holzbaur et al.* 1996, 66). Furthermore, the catalytic converter increases fuel use and thus CO_2 emissions (*Cole* 2000b, 76) and it consists in part of platinum, a stock pollutant which starts to accumulate in the soil. Second, end-of-pipe technologies have been the most important determinant in causing the downturn of the EKC for the flow pollutant SO_2 (*de Bruyn* 1997, 493 and *Ekins* 1997, 822). To run an end-of-pipe technology, however, energy is needed. Since most energy is obtained by burning fossil fuels, this causes higher CO_2 emissions.

More generally, every end-of-pipe technology consumes energy and therefore increases CO_2 emissions (*Faber et al.* 1996, 272). Thus as long as the abated pollutant is a flow pollutant, every end-of-pipe technology is an example for an abatement technology of a flow pollutant with an external effect on a stock pollutant. So *Stern* (1998, 183–184) has not only noticed a shift from SO_2, but also from NO_x emissions to CO_2 emissions. If inputs are treated (by using energy) such that less flow pollution is emitted in the production process – for instance, the cleaning of high sulphur coal (*Selden et al.* 1999) – then the technology used for this treatment has the same effect as an end-of-pipe technology: It reduces emissions of a flow pollutant but increases CO_2-emissions.

While the negative external effect of abatement of flow pollution has almost no influence on the solution of myopic governments as derived in this chapter, it becomes important in chapter 9 where we analyse the solution of far-sighted governments. To my knowledge the model developed in part III of this book is the first model in which abatement of one pollutant has an external effect on another pollutant.

The main interest of part III of this book lies in the myopic solution which is analysed in this chapter whereas we will only sketch the far-sighted solution in chapter 9. There are good reasons to focus on the myopic solution: It is impossible to internalize all environmental damages, first, because of coordination problems and transaction costs (*Smulders* 2000, 641–642) which make it simply infeasible to closely monitor all (potential) pollutants[2] and, second, because some damages are difficult to measure, in particular soil erosion (*van Kooten* and *Bulte* 2000, 387), desertification, pollution and depletion of groundwater aquifers (*Cole* 1999, 95), biodiversity loss (*Dietz* 2000, 3), acidification (*Neumayer* 1998, 168), extinction of animal and plant species, climate change, and the risk of nuclear power stations. So there are always unregulated pollutants.[3] These pollutants tend to be stock pollutants since the incentives to monitor them are smaller than for flow pollutants. Hence, the myopic solution of our model might be closer to reality than the far-sighted one. In fact, the myopic solution is consistent with the empirical evidence.

Abatement of flow, but not of stock pollutants might also be the outcome of a lobbying process: To pacify the 'green' lobby the government cannot be inactive. To loose as little votes as possible the government abates flow pollutants because the improvements can immediately be discerned while the abatement of stock pollutants would not be perceived by the uninformed electorate.

In part III we concentrate attention on the substitution of a flow pollutant by a stock pollutant. As argued in section 3.3, however, it is also possible that abatement of local pollutants aggravates the problem of transboundary or global pollutants. The internalization of a currently existing externality might even result in emissions of a hitherto not emitted chemical substance due to technological progress. Furthermore, substitution might also be indirect: Progress in technological efficiency does not solve all environmental problems as is often thought, but instead typically leads to higher consumption of natural resources since the gains are

---

[2]We will never be able to determine all adverse effects that synthetic chemicals – pure and in mixture – have on the environment because the costs would be prohibitive and because it is impossible to analyse the effects on all species since we do not even know all species (*Huesemann* 2001, 274).

[3]If the government regulates only some pollutants, but leaves other pollutants unregulated, the firms substitute away from the regulated to the unregulated pollutants (*Devlin* and *Grafton* 1994).

reinvested in other polluting activities according to *Wackernagel* and *Rees* (1997a, 173 and 1997b, 18–20). Energy gained from fossil fuels might also be replaced by nuclear power – an energy source with its own problems (*Scruggs* 1998, 269–271). Finally, substitution might not only be between pollutants, but also between countries (cf. section 3.7): The polluting firms in high-income countries might just migrate to lower income countries.

So far most researchers sidestepped the problem of substitution by looking at some sort of aggregate pollution (cf. paragraph 2.3.2.1). However, aggregation is difficult as the weights of different pollutants in the aggregate are always somewhat arbitrary. Although we also aggregate pollution in this chapter, we focus on the PIR of the single pollutants, not on the behaviour of the aggregate.

This chapter is organized as follows. Section 8.2 presents the assumptions of the model. In section 8.3 the solution of a myopic government in a given period is determined. In section 8.4 we use this solution to derive a phase diagram with which we examine the optimal path followed by of a succession of myopic governments in section 8.5. Two important consequences of this path are discussed in section 8.6. Section 8.7 concludes.

8.2 The model

To analyse the interaction of stock and flow pollutants an overlapping generations model is used. The model is an extension to two pollutants of the model of *John* and *Pecchenino* (1994). Every period a new generation is born and lives for two periods. The population size of each generation is assumed to be constant at L. In each period there are people of two different generations: The young and the old. While only the young are working and get a wage in return, only the old derive utility from consumption c and suffer from pollution P (small letters indicate per capita values).[4] Utility u of the representative consumer is given by $u(c, P)$ for each generation where $u_c > 0$ and $u_P < 0$ as before. As we will see each generation solves a static problem (where the solution determines the initial condition of the next generation). Hence, we can apply our method for static models developed in section 4.2: We assume that consumption and environmental quality (or $-P$) are both normal goods. So the derivatives of the marginal rate of substitution, $MRS = -u_c/u_P > 0$, are both negative

$$
\begin{aligned}
MRS_c &= \frac{-u_{cc}u_P + u_{Pc}u_c}{u_P^2} < 0 \\
MRS_P &= \frac{-u_{cP}u_P + u_{PP}u_c}{u_P^2} < 0
\end{aligned}
$$

(see (4.1) and (4.2)). In contrast, *John* and *Pecchenino* (1994) make stronger, cardinal assumptions ($u_{cc} < 0$, $u_{PP} < 0$, and $u_{cP} \leq 0$) which are not necessary to derive our results. Since it is argued that pollution as perceived by the general public is an aggregate measure (*Wu* 1998), we assume that pollution P is the sum of the two pollutants P^1 and P^2, $P = P^1 + \lambda P^2$ where λ is a constant describing how harmful the two pollutants are. If $\lambda > 1$, P^2 is more harmful and if $\lambda < 1$, P^1 has more damaging effects.

Competitive firms produce output Y using capital K and labour L in a Cobb-Douglas production function with constant returns to scale. Thus in period t we have $Y_t = F(K_t, L_t) = \beta K_t^{\alpha} L_t^{1-\alpha}$ where $0 < \alpha < 1$. Dividing the production function by L we obtain $y_t = f(k_t) = \beta k_t^{\alpha}$ where $y = Y/L$ and $k = K/L$ are per capita values.[5] Firms maximize their profits (where the

[4] Following *John* and *Pecchenino* (1994) we therefore simplify by concentrating on the choice between investment in pollution abatement and investment in physical capital (see below) and by abstracting from the consumption-saving choice.

[5] If we did not assume a Cobb-Douglas production function, we would have to assume $f_k > 0$, $f_{kk} < 0$, $f_{kkk} > 0$, $f_k + kf_{kk} > 0$, $f_{kk} + kf_{kkk} > 0$,

consumption good is the numeraire good)

$$max_{K,L} \quad \beta K_t^{\alpha} L_t^{1-\alpha} - w_t L_t - r_t K_t - \delta K_t$$

taking as given the wage w, the interest rate r, and the depreciation rate of capital δ. Solving this problem we find

$$r_t = \alpha\beta K_t^{\alpha-1} L_t^{1-\alpha} - \delta = \alpha\beta k_t^{\alpha-1} - \delta \tag{8.1}$$
$$w_t = (1-\alpha)\beta K_t^{\alpha} L_t^{-\alpha} = (1-\alpha)\beta k_t^{\alpha}. \tag{8.2}$$

As labour is supplied inelastically by the young, the wage adjusts to ensure equilibrium in the labour market (*Ihori* 1978). The capital market is discussed below.

Pollutant P^1 is a flow pollutant. The flow of P^1 in period t is given by[6]

$$P_t^1 = g(k_t, a_t^1)$$

---

and $2f_{kk} + kf_{kkk} < 0$. Therefore we use the simple Cobb-Douglas function which fulfils all these conditions and additionally allows to explicitly derive two further results in (8.14) and footnote 17 below.

[6]Here we deviate substantially from *John* and *Pecchenino* (1994). First, these authors take environmental quality instead of pollution. The finding of an EKC in their model might therefore suggest that the EKC applies to environmental quality generally – an interpretation which is wrong as shown in chapters 2 and 3. Second, the natural level of environmental quality in the model of *John* and *Pecchenino* (1994) is zero. It is unclear what a positive level of environmental quality, i.e. a higher level than the natural one, should be. *John* and *Pecchenino* (1994) write that environmental quality might be interpreted as the inverse of the concentration of CFCs. Then a natural level of zero means that the natural concentration of CFCs is infinity (!), i.e. the worst possible level. In the model of *John et al.* (1995) the natural level of environmental quality – which is again zero – is also the worst possible level because it is assumed that environmental quality is always positive! In contrast, we assume that pollution is always nonnegative. Third, in our model pollution is generated by capital, whereas in the model of *John* and *Pecchenino* (1994) it is generated by consumption. Fourth, *John* and *Pecchenino* (1994) assume that environmental quality depends linearly on consumption and abatement, contrary to the literature where it is assumed that $P_{cc} > 0$ (or $P_{kk} > 0$) and $P_{aa} > 0$ (*Forster* 1973, *Gruver* 1976, *Selden* and *Song* 1995, *McConnell* 1997, *Ansuategi et al.* 1998, and *Ansuategi* and *Perrings* 2000).

where a^1 are the abatement expenditures used to abate the flow pollutant P^1. We assume $g_k > 0$, $g_{kk} \geq 0$, $g_a < 0$, $g_{aa} > 0$, and $g_{ak} = g_{ka} \leq 0$. These are the same assumptions than we made for $P(K, A)$ in chapter 7 (see section 7.2). So contrary to *John* and *Pecchenino* (1994) and contrary to many other researchers (see footnote 3 in chapter 5) which for simplicity assume $g_{ak} = 0$, we assume $g_{ak} \leq 0$ which is more general and plausible: With more capital (more emissions) abatement expenditures are more efficient (cf. section 5.2). A further assumption is that $\lim_{a^1 \to 0} |g_a(\cdot, a^1)| < \infty$. If capital (emissions) is equal to zero, there is no pollution, $g(0, \cdot) = 0$. Finally, $P^1 \geq 0$ must always hold: We assume $\lim_{a^1 \to \infty} g(\cdot, a^1) = 0$. This is not unreasonable, since the marginal costs of reducing pollution rise steeply as pollution tends towards zero (*Neumayer* 1998, 166).

The second pollutant is a stock pollutant. The stock of P^2 in period t is given by

$$P_t^2 = \theta P_{t-1}^2 + h(k_{t-1} + \gamma a_{t-1}^1, a_{t-1}^2) \tag{8.3}$$

where a^2 are the abatement expenditures used to abate the stock pollutant P^2. Nature assimilates a constant share of last period's stock of P^2, $(1 - \theta)P_{t-1}^2$ where $0 < \theta < 1$. But the stock of P^2 increases because of emissions h. Capital k and – as discussed in the introduction – possibly abatement of the flow pollutant a^1 have a negative external effect on P^2: $\partial h / \partial (k + \gamma a^1) =: h_k > 0$ where $\gamma \geq 0$. We assume that the external effect of a^1 is smaller than that of k, i.e. $\gamma \leq 1$. Emissions of P^2 can also be abated, $\partial h / \partial a^2 =: h_a < 0$.[7] The negative external effect of emissions h are only felt one period later because P^2 needs time to accumulate and because the damage needs time to materialize.[8] So P_t^2 is given in period

[7]Although it is often stated that in the case of CO_2 available abatement technologies are prohibitively expensive (see for example *Nentwig* 1995, 231 and *Vogel* 1999, 126), these technologies do exist. The costs of depositing CO_2 underground or in the deep sea are even reasonably low (*Herzog et al.* 2000).

[8]For CO_2 estimates of the time lag between increases in radiative inputs and the climate change range from 6 to 95 years (*Nordhaus* 1991, 922).

t and cannot be altered by actions taken in period t. Analogous to the assumption for P^1 we assume $\lim_{a^2 \to 0} |h_a(\cdot, a^2)| < \infty$ and $h(0, \cdot) = 0$.

Following *John et al.* (1995) we assume a government which collects a per capita lump sum tax τ_{t-1} on the wage of the young. The government uses the tax revenues to construct abatement technologies for P^1 and P^2 which are used in the next period, $\tau_{t-1} = a_t^1 + a_t^2$. To simplify we assume that at the end of period t the abatement technologies are fully worn out, i.e. the depreciation rate for the abatement technology is 100%. The young give all their after tax wages, their savings $s_{t-1} = w_{t-1} - \tau_{t-1}$, to the firms. Since consumers give their savings inelastically to the firms, the interest rate r (given in (8.1)) must adjust to ensure market clearing, $s_{t-1} = k_t$. In the next period the firms pay their capital stock, including interest rates, back to the now old who use their savings for consumption, $c_t = (1 + r_t)s_{t-1}$. The firms receive new capital from the next generation.

8.3 The solution in a given period

In this chapter we analyse how the economy develops when in each period a myopic government is in office. In chapter 9 we will give some insights into the solution of far-sighted governments. In setting the tax rate, τ_{t-1}, and choosing a_t^1 and a_t^2 the myopic government of period $t-1$ maximizes the utility of all people living in period $t-1$. However, the utility of the old is not influenced by this decision as capital and abatement only change in the next period. So the government actually maximizes the utility that the young of period $t-1$ will derive in period t when they are old. Another possibility for setting the wage tax is that there is a vote about how high it should be. Then the old of period $t-1$ do not vote. Thus it is assumed that all intra-generational externalities are internalized whereas the inter-generational externalities are not. Since P_t^2 is given for generation t, this model analyses what can happen when the externality of one pollutant is internalized and the externality of another is not. The government's problem in period $t-1$ is to maximize the utility of the generation which is old in period t,

$$\max_{a_t^1, a_t^2} \quad u(c_t, P_t^1 + \lambda P_t^2)$$

$$\begin{aligned}
\text{subject to} \qquad & \tau_{t-1} = a_t^1 + a_t^2 && (8.4)\\
& s_{t-1} = w_{t-1} - \tau_{t-1} && (8.5)\\
& k_t = s_{t-1} && (8.6)\\
& r_t = \alpha\beta k_t^{\alpha-1} - \delta && (8.7)\\
& c_t = (1 + r_t)s_{t-1} && (8.8)\\
& P_t^1 = g(k_t, a_t^1) && (8.9)\\
& a_t^1 \geq 0, \;\; a_t^2 \geq 0 && (8.10)
\end{aligned}$$

taking as given w_{t-1} and P_t^2 since w_{t-1} is determined by k_{t-1} (see (8.2)) and since P_t^2 cannot be influenced any more (see (8.3)). Note finally that feasibility requires

$$f(k_t) = \delta k_t + c_t + a_{t+1}^1 + a_{t+1}^2 + (k_{t+1} - k_t),$$

i.e. output $f(k_t)$ is used to finance capital depreciation, δk_t, consumption in this period, c_t, abatement in the next period, $a_{t+1}^1 + a_{t+1}^2$, and growth of the capital stock, $k_{t+1} - k_t$.

We now study the solution in a given period, i.e. the solution the government chooses given P_t^2 and k_{t-1}. Since a_t^2 does not show up in the utility function and since it competes with a_t^1 and k_t for the use of w_{t-1} ($w_{t-1} = a_t^1 + a_t^2 + k_t$), it is clear that $a_t^2 = 0$: The myopic government in period $t-1$ never abates the stock pollutant because this has only an effect on P^2 in period $t+1$ which the myopic government neglects. Instead it abates the flow pollutant which has an immediate effect or it lets capital grow which increases consumption: Inserting (8.6) and (8.7) into (8.8) we can write consumption as a function of the capital stock,

$$\begin{aligned} c_t = c(k_t) &= (1-\delta)k_t + \alpha\beta k_t^{\alpha}, \\ c_k = dc/dk &= 1-\delta+\alpha^2\beta k_t^{\alpha-1} > 0, \\ c_{kk} &= \alpha^2(\alpha-1)\beta k_t^{\alpha-2} < 0. \end{aligned} \tag{8.11}$$

Setting $a_t^2 = 0$ it follows that maximizing with respect to a_t^1 is equivalent to maximizing with respect to k_t because $w_{t-1} = a_t^1 + k_t$ where w_{t-1} is given. Inserting (8.4) – (8.9) and (8.11) into the utility function, the problem of the myopic government simplifies to

$$\max_{k_t} \; u[c(k_t), g(k_t, w_{t-1} - k_t) + \lambda P_t^2] + \phi(w_{t-1} - k_t)$$

where ϕ is the multiplier of the nonnegativity constraint $a_t^1 \geq 0$. Dividing the first order condition by $-u_P$ we find

$$MRS \cdot c_k = g_k - g_a - \phi/u_P. \tag{8.12}$$

There is a simple graphical interpretation of (8.12) in the P_t - k_t space of figure 8.1: On the one hand, the *consumption possibilities curve CPC* shows all feasible combinations of k_t and P_t. The CPC is thus defined by

$$P_t = g(k_t, w_{t-1} - k_t) + \lambda P_t^2. \tag{8.13}$$

The government chooses $k_t \in [0, w_{t-1}]$. One extreme case is the maximal possible level $k_t = w_{t-1}$ as shown in figure 8.1. According to (8.13) pollution P_t is then $g(w_{t-1}, 0) + \lambda P_t^2$. The other extreme case is to use all the wage for abatement such that $k_t = 0$ and $P_t = \lambda P_t^2$ since $g(0, a^1) = 0$. In between these two extremes the CPC is rising and convex because (8.13) shows that $P_k = g_k - g_a > 0$ and $P_{kk} = g_{kk} - 2g_{ak} + g_{aa} > 0$. On the other hand, we can draw an indifference curve into figure 8.1. The indifference curves are rising and concave (from $u_c c_k dk + u_P dP = 0$ we derive $(dP/dk)|_{u\ constant} = MRSc_k > 0$ and $(d(MRSc_k)/dk)|_{u\ constant} = MRS_c c_k^2 + MRSc_{kk} + MRS_P c_k (dP/dk)|_{u\ constant} < 0$). Hence,

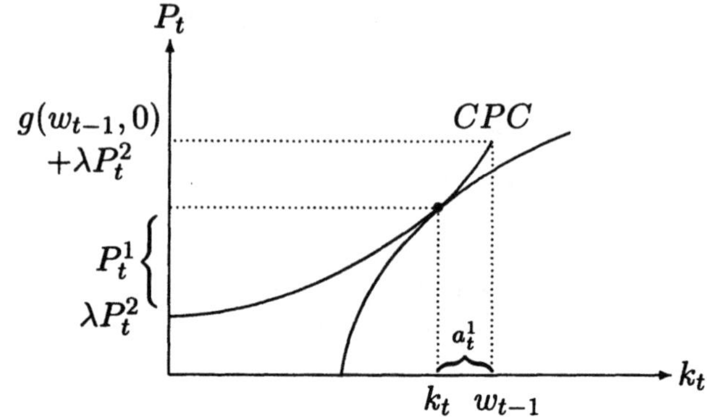

Figure 8.1: The solution in period t.

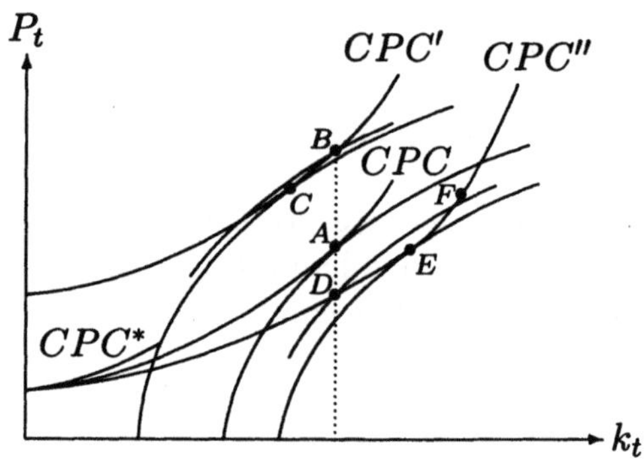

Figure 8.2: Solution for higher P_t^2 ('), higher k_{t-1} ("), or lower k_{t-1} (*).

(8.12) tells us that at an interior solution (with $a^1 > 0$ and $\phi = 0$) the convex CPC is tangent to a concave indifference curve. Thus the second order condition for a maximum is satisfied.

Next we analyse how the solution changes either when stock pollution P_t^2 is higher or when last period's capital stock k_{t-1} (and thus last period's wage w_{t-1}) is higher. For an interior solution we find the following lemma:

Lemma 8.1 *If P_t^2 rises, the capital stock, k_t, falls, but abatement expenditures, a_t^1, grow in the optimum. If k_{t-1} rises, the optimal level of k_t and a_t^1 both increase.*

A mathematical proof is given in the appendix. But this result can also be derived graphically following figure 8.2. Note first that the slope of the indifference curve becomes smaller with rising pollution and with rising capital ($d(MRSc_k)/dP = MRS_{P}c_k < 0$ and $d(MRSc_k)/dk = MRS_c c_k^2 + MRSc_{kk} < 0$). Using this result we can prove lemma 8.1: On the one hand, if P_t^2 rises (holding k_{t-1} constant), the CPC shifts parallel upwards (CPC' in figure 8.2) because the intercept of the CPC, λP_t^2, increases (see (8.13)). At point B in figure 8.2, i.e. on the CPC' at the same capital stock as before, the indifference curve is flatter than at point A, but the CPC has the same slope at points A and B. Thus the new optimal point C lies at a smaller capital stock. Since $a_t^1 + k_t$ equals w_{t-1}, a smaller k_t implies a higher a_t^1. Intuitively, as P_t^2 rises, the marginal damage of pollution rises so that we abate more.

On the other hand, if k_{t-1} and thus w_{t-1} rises (holding P_t^2 constant), more a_t^1 is possible at the same level of k_t. Thus P_t^1 and P_t are lower and the CPC shifts downwards (CPC'' in figure 8.2). Furthermore, the endpoint of the CPC (where $k_t = w_{t-1}$) shifts to the right and upwards since the higher wage increases the capital stock and therefore also P_t^1 and P_t. At point D in figure 8.2, i.e. on the CPC'' at the same capital stock as before, the indifference curve is steeper than at point A and the CPC'' is flatter (as $d(g_k(k, a^1) - g_a(k, a^1))/da^1 = g_{ak} - g_{aa} < 0$). So the new optimal point E lies at a higher capital stock. If a_t^1 stayed constant, we would be at point F. Since the indifference curve is flatter at point F than at point A (as k_t and P_t are higher) and since the CPC is steeper ($d(g_k(k, a^1) - g_a(k, a^1))/dk = g_{kk} - g_{ak} > 0$), the new optimal point E lies at a higher level of a_t^1. Intuitively, capital and abatement rise with the rising wage since consumption and environmental quality (or $-P$) are both normal goods.

8.4 Derivation of the phase diagram

For every given level of k_{t-1} and P_t^2 we can therefore determine the optimal level of k_t and a_t^1 with figure 8.2. Then (8.3) gives the stock of P_{t+1}^2. Hence, using a phase diagram in the P_t^2 - k_{t-1} space we can analyse how the economy evolves over time when the government of each period $t-1$ myopically maximizes the utility of period t. In this section we derive the phase diagram.

Until now we have assumed that abatement expenditures are positive. However, the economy can be so poor and endowed with such a small stock of P^2 inherited from earlier generations that it is not worth to abate any flow pollution P^1 as on the CPC^* in figure 8.2. Thus we can find a line in the phase diagram – the $a_t^1 = 0$-line – below which abatement expenditures are optimally chosen to be zero.

Lemma 8.2 *Abatement expenditures, a_t^1, are zero at low levels of k_{t-1} and of P_t^2, i.e. below the $a_t^1 = 0$-line. The $a_t^1 = 0$-line is downward sloping (see figure 8.3*[9]*).*

The proof is provided in the appendix. Graphically, for a point on the $a_t^1 = 0$-line the CPC is tangent to an indifference curve at the endpoint of the CPC. The $a_t^1 = 0$-line is downward sloping because for a lower stock of P_t^2 we need higher k_{t-1} to have again a tangency point at the endpoint of the CPC. Intuitively, with a higher stock of pollution abatement is more attractive and thus starts to increase at a lower level of capital.

To analyse the myopically optimal path in the phase diagram we now derive the $\bar{k}$-line on which capital stays constant, $k_t = k_{t-1}$, and the $\bar{P}^2$-line on which the stock of P^2 stays constant, $P_{t+1}^2 = P_t^2$. To characterize the $\bar{k}$-line we need an additional assumption:

[9]The curvature, i.e. convexity or concavity, of the lines in figures 8.3–8.5 are arbitrarily chosen. All that counts is whether the lines are rising or falling.

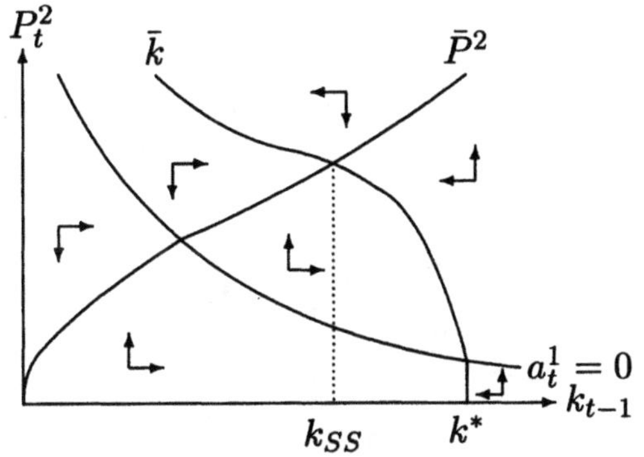

Figure 8.3: The phase diagram.

Assumption 8.1 *Along the $\bar{k}$-line $\partial k_t/\partial k_{t-1} < 1$ is satisfied. In other words,*

$$\begin{aligned} X \;:=\; & MRS_c c_k^2 + MRS c_{kk} + MRS_P c_k g_k - g_{kk} + g_{ak} \\ & +(1 - w_{k_{t-1}})(g_{ak} - g_{aa} - MRS_P c_k g_a) < 0 \end{aligned}$$

holds where $w_{k_{t-1}} = dw_{t-1}/dk_{t-1}$.

Note that assumption 8.1 is surely satisfied if $1 - w_{k_{t-1}} \geq 0$. This is the case for great k. In the proof of lemma 8.3 we show that assumption 8.1 is not very restrictive.

Lemma 8.3 1. *Below the $a_t^1 = 0$-line the $\bar{k}$-line is vertical at k^* where*

$$k^* = [(1-\alpha)\beta]^{\frac{1}{1-\alpha}}. \tag{8.14}$$

Above the $a_t^1 = 0$-line the $\bar{k}$-line is negatively sloped if assumption 8.1 holds. To the left (right) of the $\bar{k}$-line capital rises (falls) as indicated by the arrows in figure 8.3.

2. *The $\bar{P}^2$-line is positively sloped, but becomes flatter when it crosses the $a_t^1 = 0$-line from below. To the left (right) of the $\bar{P}^2$-line P^2 falls (rises) as shown in figure 8.3.*

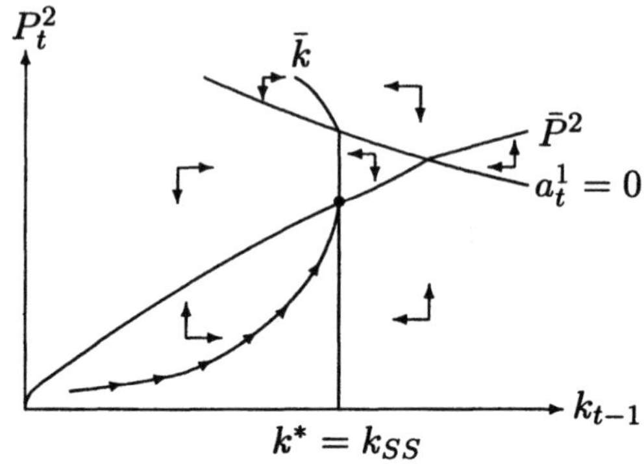

Figure 8.4: The phase diagram with zero abatement at the steady state.

The proof is relegated to the appendix. The $\bar{k}$-line is downward sloping because k_t rises less than proportionally with k_{t-1} according to assumption 8.1. So P_t^2 must fall to ensure that k_t increases at the same speed as k_{t-1}. The $\bar{P}^2$-line is upward sloping because with a higher stock of P_t^2 nature assimilates more $((1-\theta)P_t^2)$ and thus emissions generated by capital can be higher. There is exactly one intersection of the $\bar{k}$- and the $\bar{P}^2$-line, i.e. exactly one steady state. The steady state capital stock is k_{SS}. We will not analyse the steady state more closely since it will turn out that the steady state is not important for our argument.

8.5 The equilibrium path of myopic governments

Let us now analyse what the equilibrium path of a succession of myopic governments looks like. We start with small levels of capital and of the stock pollutant because this is the interesting case in which an EKC might emerge. Then abatement expenditures are first set equal to zero. It is even possible that abatement expenditures are still zero at the steady state as shown in figure

8.4. As long as they are zero, capital growth is independent of the stock of P^2 and capital asymptotically approaches the stable k^* (see proof of lemma 8.3.1). As shown in figure 8.4 the stock pollutant P^2 grows along this path. The same is true for the flow pollutant since $P^1 = g(k, 0)$ increases with k.

This result corresponds to the case in which no abatement technology is available. For the rest of this chapter we therefore turn to the more interesting case of an interior steady state with abatement. The optimal path is depicted in the lower panel of figure 8.5. Since the main purpose of this model lies in the explanation of the empirical evidence, we only analyse the path until the maximal capital stock is reached. Considering this part of the path we can now prove one of the main results of this chapter:

Proposition 8.1 *The flow pollutant P^1 first rises and then falls: There is an EKC for P^1. The stock pollutant P^2, however, rises throughout.*

Proof. The proof involves three steps. First, we examine how a^1 behaves along the myopically optimal path. Second, we show that capital overshoots its steady state stock and that P^2 rises. Third, we analyse the behaviour of P^1 along the optimal path.

First, at an interior solution abatement expenditures change along the optimal path according to

$$\Delta a_t^1 = \frac{\partial a_t^1}{\partial k_{t-1}} \Delta k_{t-1} + \frac{\partial a_t^1}{\partial P_t^2} \Delta P_t^2$$

where $\Delta x_t = x_t - x_{t-1}$ and where $\partial a_t^1 / \partial k_{t-1} > 0$ and $\partial a_t^1 / \partial P_t^2 > 0$ are derived in the proof of lemma 8.1. Thus abatement expenditures are rising after the optimal path has crossed the $a_t^1 = 0$-line and before the $\bar{k}$-line is reached because there both variables are rising, i.e. $\Delta k_{t-1} > 0$ and $\Delta P_t^2 > 0$. When the economy crosses the $\bar{k}$-line, $\Delta k_{t-1} = 0$ and $\Delta P_t^2 > 0$ hold such that a^1 rises.

Second, capital overshoots since the stock pollutant P^2 needs time to accumulate: Suppose that P^2 is in an equilibrium in

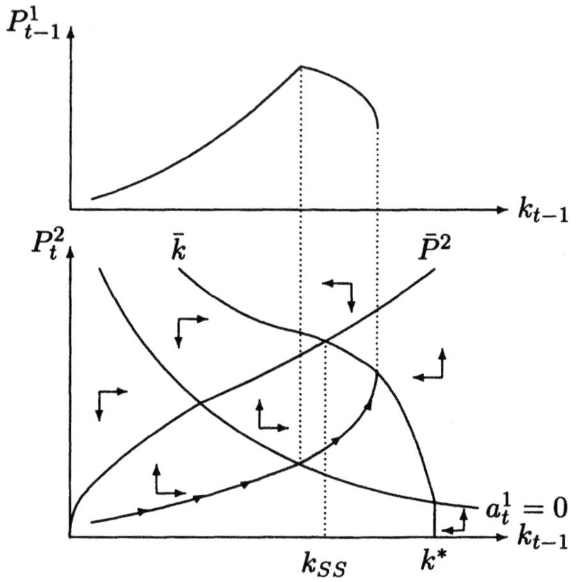

Figure 8.5: The myopically optimal path of the economy.

which the assimilated stock, $(1-\theta)P_t^2$, equals newly generated emissions, $h(k_t + \gamma a_t^1, 0)$ (see (8.3)). If emissions then suddenly increased and stayed at their higher level forever, P^2 would immediately start to rise, but would only approach its new equilibrium level asymptotically.[10] Since capital and – as we have shown – abatement expenditures are both growing along the optimal path, emissions 'suddenly' increase in every period. So P^2 does not have the time to accumulate. Therefore P^2 is still smaller than its steady state stock when the economy reaches its steady state

[10]Suppose that emissions rise to $\epsilon := h(k + \gamma a^1, 0)$ in period T. The new equilibrium value of P^2 is $\bar{P}^2 = \epsilon/(1-\theta)$. In period T we know that $P_T^2 = \pi \bar{P}^2$ where $\pi < 1$. So the difference to the equilibrium is $(1-\pi)\bar{P}^2$. According to (8.3) next periods stock is given by $P_{T+1}^2 = \theta\pi\epsilon/(1-\theta) + \epsilon = (1-\theta+\theta\pi)\epsilon/(1-\theta)$ and the difference is now $\bar{P}^2 - P_{T+1}^2 = \theta(1-\pi)\epsilon/(1-\theta) = \theta(1-\pi)\bar{P}^2$. Therefore the difference to the equilibrium declines to a constant fraction θ of the difference in the last period.

capital stock, k_{SS}. It follows from the phase diagram that capital continues to grow: Capital overshoots. It reaches its maximal value when it crosses the $\bar{k}$-line. Hence, P^2 rises along the optimal path until the maximal capital stock is reached.

Third, it remains to be discussed how P^1 changes along the optimal path. As long as abatement expenditures are zero, $P^1 = g(k, 0)$ increases with capital. For an interior solution we use $P_t^1 = g(k_t, a_t^1)$ and (8.19) and (8.20) from the proof of lemma 8.1 to derive

$$\begin{aligned}\Delta P_t^1 &= g_k \tfrac{\partial k_t}{\partial k_{t-1}} \Delta k_{t-1} + g_a \tfrac{\partial a_t^1}{\partial k_{t-1}} \Delta k_{t-1} + g_k \tfrac{\partial k_t}{\partial P_t^2} \Delta P_t^2 + g_a \tfrac{\partial a_t^1}{\partial P_t^2} \Delta P_t^2 \\ &= \tfrac{-w_{k_{t-1}}[g_k(g_{aa}-g_{ak})+g_a(g_{kk}-g_{ak})-g_a(MRS_c c_k^2+MRS c_{kk})]}{MRS_c c_k^2+MRS c_{kk}+MRS_P c_k(g_k-g_a)-g_{kk}+2g_{ak}-g_{aa}} \Delta k_{t-1} \\ &\quad + \tfrac{(g_a-g_k)\lambda MRS_P c_k}{MRS_c c_k^2+MRS c_{kk}+MRS_P c_k(g_k-g_a)-g_{kk}+2g_{ak}-g_{aa}} \Delta P_t^2.\end{aligned} \tag{8.15}$$

The fraction in the second line of (8.15) is negative. The fraction in the first line of (8.15) is negative if $g_k(g_{aa} - g_{ak}) + g_a(g_{kk} - g_{ak}) - g_a(MRS_c c_k^2 + MRS c_{kk}) < 0$ or if

$$- MRS_c c_k^2 - MRS c_{kk} > \frac{g_{kk} g_a - g_{ak} g_k - g_{ak} g_a + g_{aa} g_k}{-g_a}. \tag{8.16}$$

This has a striking similarity to (5.8) on page 121: The left hand side of (8.16) or (5.8) shows by how much the slope of the indifference curve falls when we move horizontally to the right in figure 8.2 (see section 8.3) or in figure 5.2 on page 118. The right hand side of (8.16) or (5.8) shows by how much the slope of the CPC falls holding constant P_t^1 in chapter 8 or P in chapter 5 (cf. section 5.2 and subsection 5.3.2). A horizontal movement to the right in figure 8.2 which holds P_t^1 constant implies that P_t^2 remains also constant. Thus the movement of the CPC is caused by rising k_{t-1} (cf. (8.15)). If (8.16) or (5.8) is satisfied (violated), the slope of the indifference curve falls faster (slowlier) than the slope of the CPC. Therefore the new optimal point lies at a smaller (higher) pollution level (cf. subsection 5.3.2). In chapter 5 we have shown that (5.8) can be satisfied or violated. Hence, the same is true for (8.16).

On the one hand, if (8.16) is satisfied, P^1 falls at an interior solution because Δk_{t-1} and ΔP_t^2 are both positive before the path reaches the $\bar{k}$-line in the lower panel of figure 8.5. Hence, below the $a_t^1 = 0$-line P^1 increases along the optimal path, but as soon as abatement expenditures become positive, P^1 declines as shown in the upper panel of figure 8.5. The EKC lies in the P^1 - GDP space. Since GDP $= f(k)$ increases with capital, the inverted U-shaped path in the P^1 - k space also implies an EKC for the flow pollutant.[11]

On the other hand, if (8.16) is violated, P^1 falls along the path if the path is relatively steep after crossing the $a_t^1 = 0$-line, i.e. it follows from (8.15) that $\Delta P^1 \leq 0$ if

$$\frac{\Delta P_t^2}{\Delta k_{t-1}} \geq \frac{w_{k_{t-1}}[g_k(g_{aa}-g_{ak})+g_a(g_{kk}-g_{ak})-g_a(MRS_c c_k^2+MRSc_{kk})]}{(g_a-g_k)\lambda MRS_P c_k}. \quad (8.17)$$

This condition is surely satisfied before the path reaches the $\bar{k}$-line where the slope of the path becomes infinity. Thus even if the path is relatively flat ((8.17) violated) directly after crossing the $a_t^1 = 0$-line such that P^1 continues to increase, P^1 starts to fall before the $\bar{k}$-line is reached.[12] Hence, we also find an EKC. □

The main driving force behind the EKC for P^1 is that we first are at a corner solution with zero abatement expenditures, but later on abatement increases. Since myopic governments only abate P^1 and since P^2 always increases, a^1 rises so fast that $P^1 = g(k, a^1)$ declines although capital grows. The result of proposition 8.1 – an EKC for the flow pollutant but a monotonically rising PIR for the stock pollutant – is consistent with the empirical evidence.

[11]This holds also true if we use income $= f(k) - \delta k$ instead of GDP (see footnote 17 below and note that $d(\text{income})/dk = f_k - \delta = r$).

[12]This is a continuous time argument which might be wrong in our discrete time model. However, since the model is aimed at explaining the empirical evidence, we do not give any weight to this theoretical possibility.

Furthermore, from the proof of proposition 8.1 we can draw two conclusions which are presented in the next section.

8.6 Two important consequences

As mentioned in chapter 1 it is often concluded from the finding of an EKC for a few pollutants that all other pollutants also follow an EKC. In our model, however, the EKC for P^1 is accompanied by a monotonically rising PIR for P^2. Hence, this conclusion is clearly wrong in our model. Even worse, if we assume that there is no stock pollutant or – put differently – that the stock 'pollutant' is harmless, i.e. if $\lambda = 0$, there might not be an EKC for P^1:

Proposition 8.2 *Suppose that (8.16) is violated along the optimal path, i.e. that a higher* k_{t-1} *increases* P_t^1*. Then the PIR for* P^1 *is monotonically rising if the stock pollutant is harmless, i.e. if* $\lambda = 0$.

Proof. If $\lambda = 0$, the second line in (8.15) drops out.[13] So P^1 is monotonically increasing if (8.16) is violated. □

If there is another, ever rising pollutant which cannot be abated in the short-run ($\lambda > 0$), an EKC for P^1 arises as shown in proposition 8.1. In contrast, if there is no such pollutant ($\lambda = 0$), the PIR for P^1 is monotonically rising if (8.16) is violated. Hence, *we might only observe EKCs for certain flow pollutants because the stock of other pollutants is increasing.*

A second important result also follows from the proof of proposition 8.1:

Proposition 8.3 *Suppose that (8.16) is satisfied or that (8.16) is violated and that (8.17) holds, i.e. that the optimal path is relatively steep after crossing the* $a_t^1 = 0$*-line. Then the turning*

[13] In addition, the $a_t^1 = 0$- and the $\bar{k}$-line are both vertical (see (8.21) and (8.22) in the proofs of lemma 8.2 and 8.3.1). Thus capital does no longer overshoot its steady state stock.

point of the EKC for P^1 lies at lower levels of income and of P^1, if one of the following conditions is satisfied where we always hold all other variables constant: If the stock of P^2 is higher, if P^2 is more harmful (higher λ), if production is more P^1-intensive ($g(k,0)$ and $g_k(k,0)$ higher), if abatement is cheaper ($|g_a(k,0)|$ higher), if preferences are 'greener' (MRS smaller), and if consumption is higher ($c(k)$ higher), but rises slowlier with capital (c_k smaller).

Proof. If (8.16) is satisfied or if (8.16) is violated and (8.17) holds, P^1 starts declining as soon as abatement expenditures become positive. Thus if the optimal path crosses the $a_t^1 = 0$-line at a low capital stock, the turning point of the EKC occurs at a low capital stock and also at a low level of $P^1 = g(k,0)$. If P^2 is high, i.e. if the initial stock of P^2 is high and if P^2 rises fast with capital for $a^1 = 0$ (due to high θ and high $h(k,0)$, see (8.3)), the economy reaches the $a_t^1 = 0$-line at a smaller capital stock (see figure 8.5).

The optimal path also crosses the $a_t^1 = 0$-line at a small capital stock if the $a_t^1 = 0$-line lies at small levels of P^2. The $a_t^1 = 0$-line is defined by $MRS(c(k), g(k,0) + \lambda P^2)c_k = g_k(k,0) - g_a(k,0)$ (see (8.12)). Hence, for a given capital stock the $a_t^1 = 0$-line lies at low levels of P^2 if $g(k,0)$ and λ are high. Pollution P (and thus P^2 for given $g(k,0)$ and λ) is low, if the MRS is high, i.e. if g_k and $|g_a|$ are high and if c_k is low. If the MRS tends to be small due to high $c(k)$ or due to 'green' preferences (MRS small for given c, P), then P and P^2 must be small for given c_k, g_k, and g_a. □

Therefore if 'background pollution' (P^2) is higher and more harmful, the turning point of the EKC for P^1 lies at lower levels of capital and of flow pollution. Furthermore, for different pollutants P^1 or for different pollution functions g in different countries if production is P^1-intensive or if abatement is cheap, a^1 turns positive at a lower capital stock. So the model can also accommodate the observation of different turning points for different pollutants P^1 or in different countries. High consumption or 'green' preferences cause the marginal utility of consumption to be small relative to the marginal disutility of pollution. Then the turning point occurs at lower levels of capital and of P^1. Note that different countries tend to have different utility functions (*Pearson* 1994, 205) such that we expect different turning points in different countries.

In chapter 2 we have shown that most empirical studies assume that the turning point of the EKC occurs at the same income level in all countries. However, this is not an appropriate assumption because – as discussed in paragraph 2.3.3.6 – recent studies[14] allow for and find different turning points in different countries. This empirical finding is theoretically underpinned by proposition 8.3 since 'background pollution', its harmfulness,[15] the pollution function g, and the utility function all tend to be different in different countries. Further support for proposition 8.3 is supplied by *de Bruyn* (1997, 496) and *List* and *Gallet* (1999, 422): They find evidence that the turning point lies at a lower income level if pollution is high, i.e. if P^1 is high.

All the results of myopic governments would be almost the same if the abatement technology for the flow pollutant had no external effect on stock pollution, i.e. if $\gamma = 0$. The only difference for $\gamma = 0$ is that the $\bar{P}^2$-line and the optimal path are flatter after crossing the $a_t^1 = 0$-line (see (8.25) in the proof of lemma 8.3.2). Therefore the optimal path would reach the unchanged $\bar{k}$-line at a lower level of P^2 and at a higher capital stock. However, the external effect of abatement ($\gamma > 0$) becomes important in chapter 9 where we consider the path followed by far-sighted governments.

---

[14]See *Dijkgraaf* and *Vollebergh* (1998), *Stern et al.* (1998), *Koop* and *Tole* (1999), *List* and *Gallet* (1999), *Perman* and *Stern* (1999), *de Bruyn* (2000, 105–106), *Millimet* (2000), *Millimet et al.* (2000), and *Auffhammer et al.* (2001b).

[15]If P^2 is a global pollutant like CO_2, 'background pollution' is equal in all countries. However, not all countries are equally affected by global warming (λ differs across countries).

8.7 Conclusion

In this chapter we have searched for an explanation of the empirical finding that the PIR is an EKC for flow pollutants, but that the PIR is monotonically rising for stock pollutants. We have analysed an overlapping generations model with a flow and a stock pollutant. It has been shown that a succession of myopic governments follows a path on which there is an EKC for the flow pollutant, but on which the stock pollutant is ever rising. This is consistent with the empirical evidence. It is actually due to the monotonically rising stock pollutant that we surely find an EKC for the flow pollutant: Myopic governments want to keep aggregate pollution at a reasonably low level. However, since abating emissions of the stock pollutant does not have an immediate effect on the level of stock pollution, myopic governments abate flow pollution only. Because stock pollution is rising, myopic governments increase abatement expenditures for the flow pollutant. This causes the EKC. If there was no stock pollutant, in contrast, we might observe an ever rising PIR for the flow pollutant. We have also found that the turning point of the myopically optimal EKC for the flow pollutant lies at lower levels of income and of flow pollution if stock pollution is high and harmful. Similarly, the turning point lies at a small income level if flow pollution is high. This casts doubts on most empirical EKC studies because they assume that the turning point occurs at the same income level in all countries. However, it is consistent with recent empirical studies which find that the PIR differs between countries.

Hence, a succession of myopic governments always chooses an EKC for the flow pollutant because stock pollution is increasing. However, we do not yet know whether or not our assumption that the governments are myopic is crucial for this result. Therefore it is worth analysing what the solution looks like when the governments are far-sighted. We will do so in the next chapter. Thus we postpone some further important conclusions emerging from part III of the book to the end of chapter 9.

Appendix: Proofs

Proof of lemma 8.1

Totally differentiating (8.12) at an interior solution (where $\phi = 0$) yields

$$\begin{aligned}&(MRS_c c_k^2 + MRSc_{kk} + MRS_P c_k(g_k - g_a))dk_t \\ &+MRS_P c_k g_a w_{k_{t-1}} dk_{t-1} + \lambda MRS_P c_k dP_t^2 \\ &\quad = (g_{kk} - 2g_{ak} + g_{aa})dk_t + (g_{ka} - g_{aa})w_{k_{t-1}} dk_{t-1}\end{aligned} \tag{8.18}$$

where it follows from (8.2) that $w_{k_{t-1}} = dw_{t-1}/dk_{t-1} = \alpha(1-\alpha)\beta k_{t-1}^{\alpha-1} > 0$. Holding P_t^2 or k_{t-1} constant, respectively, we find

$$\begin{aligned}\frac{\partial k_t}{\partial k_{t-1}} &= \frac{w_{k_{t-1}}(g_{ak} - g_{aa} - MRS_P c_k g_a)}{MRS_c c_k^2 + MRSc_{kk} + MRS_P c_k(g_k - g_a) - g_{kk} + 2g_{ak} - g_{aa}} > 0 \\ \frac{\partial k_t}{\partial P_t^2} &= \frac{-\lambda MRS_P c_k}{MRS_c c_k^2 + MRSc_{kk} + MRS_P c_k(g_k - g_a) - g_{kk} + 2g_{ak} - g_{aa}} < 0.\end{aligned} \tag{8.19}$$

To see how a_t^1 changes with k_{t-1} and P_t^2 we totally differentiate $a_t^1 = w_{t-1} - k_t$ to obtain

$$da_t^1 = w_{k_{t-1}} dk_{t-1} - \frac{\partial k_t}{\partial k_{t-1}} dk_{t-1} - \frac{\partial k_t}{\partial P_t^2} dP_t^2.$$

So using (8.19) we derive

$$\begin{aligned}\frac{\partial a_t^1}{\partial P_t^2} &= -\frac{\partial k_t}{\partial P_t^2} = \frac{\lambda MRS_P c_k}{MRS_c c_k^2 + MRSc_{kk} + MRS_P c_k(g_k - g_a) - g_{kk} + 2g_{ak} - g_{aa}} > 0 \\ \frac{\partial a_t^1}{\partial k_{t-1}} &= \frac{w_{k_{t-1}}(MRS_c c_k^2 + MRSc_{kk} + MRS_P c_k g_k - g_{kk} + g_{ak})}{MRS_c c_k^2 + MRSc_{kk} + MRS_P c_k(g_k - g_a) - g_{kk} + 2g_{ak} - g_{aa}} > 0.\end{aligned} \tag{8.20}$$

Note that $\partial a_t^1/\partial k_{t-1} + \partial k_t/\partial k_{t-1} = w_{k_{t-1}}$. □

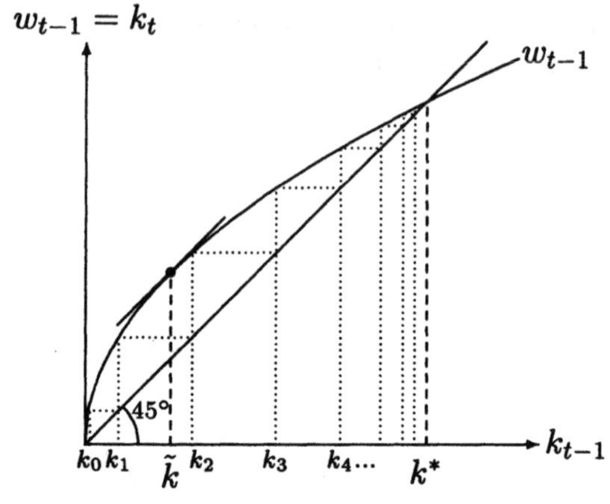

Figure 8.6: Growth of capital without abatement.

Proof of lemma 8.2

If capital goes to zero, the slope of the indifference curves, i.e. the left hand side of (8.12), goes to infinity as $\lim_{k\to 0} c_k = \infty$ (see (8.11)).[16] However, the slope of the CPC, i.e. $g_k - g_a$ on the right hand side of (8.12), is bounded since $\lim_{a^1 \to 0} |g_a| < \infty$. Therefore for small capital stocks $\phi > 0$ and $a^1 = 0$ must hold.

The $a_t^1 = 0$-line is defined by (8.12) with $a_t^1 = 0 = \phi$. Totally differentiating (8.12) yields (8.18). Inserting $dk_t = w_{k_{t-1}} dk_{t-1}$ which holds for $a_t^1 = 0$ (see (8.4) - (8.6)) we find

$$\left.\frac{dP_t^2}{dk_{t-1}}\right|_{a_t^1=0} = \frac{w_{k_{t-1}}(MRS_c c_k^2 + MRS c_{kk} + MRS_P c_k g_k - g_{kk} + g_{ak})}{-\lambda MRS_P c_k} < 0. \quad (8.21)$$

Therefore the $a_t^1 = 0$-line is downward sloping. Since $\partial a_t^1 / \partial P_t^2 > 0$ (see (8.20)), a^1 would be negative below the $a_t^1 = 0$-line. Because this is not feasible, abatement expenditures are zero below the $a_t^1 = 0$-line. □

[16] The same conclusion would also follow if we assumed that $\lim_{c\to 0} MRS = \infty$.

Proof of lemma 8.3

1. As long as abatement is zero, the government is actually doing nothing at all. The path of capital is derived from (8.6), (8.5), and (8.2) to be

$$k_t = s_{t-1} = w_{t-1} = (1-\alpha)\beta k_{t-1}^{\alpha}.$$

This is shown in figure 8.6 (the curvature of the w_{t-1}-line is derived from (8.2)). Suppose that $k_{t-1} = xk^*$ where $x > 0$ and where k^* is given in (8.14). It follows that

$$k_t = (1-\alpha)\beta(xk^*)^{\alpha} = x^{\alpha}[(1-\alpha)\beta]^{\frac{1}{1-\alpha}} = x^{\alpha}k^* = x^{\alpha-1}k_{t-1}.$$

Capital stays only constant, i.e. $k_t = k_{t-1}$, if $x = 1$, i.e. $k_{t-1} = k^*$. The $\bar{k}$-line is vertical at k^* and thus independent of P_t^2 as shown in figure 8.3. As $\alpha - 1 < 0$, it also follows that if $x < 1$, i.e. if $k_{t-1} < k^*$, capital is growing (see arrows in figure 8.3), but $k_t = x^{\alpha}k^*$ is still smaller than k^* (see figure 8.6). If $x > 1$, capital is decreasing. Thus k^* is a stable steady state which is only approached asymptotically.[17]

Turning to the interior solution ($a^1 > 0$) we totally differentiate $k_t = k_{t-1}$ which holds on the $\bar{k}$-line to find $\frac{\partial k_t}{\partial k_{t-1}} dk_{t-1} + \frac{\partial k_t}{\partial P_t^2} dP_t^2 = dk_{t-1}$. Inserting (8.19) and X from assumption 8.1 we derive

$$\left.\frac{dP_t^2}{dk_{t-1}}\right|_{k_t=k_{t-1}} = \frac{1-\partial k_t/\partial k_{t-1}}{\partial k_t/\partial P_t^2} = \frac{X}{-\lambda MRS_{Pc_k}}$$

$$= \Big(MRS_c c_k^2 + MRS c_{kk} + MRS_{Pc_k} g_k - g_{kk} + g_{ak} \qquad (8.22)$$

$$+ (1 - w_{k_{t-1}})(g_{ak} - g_{aa} - MRS_{Pc_k} g_a)\Big) \,/\, (-\lambda MRS_{Pc_k}) < 0$$

---

[17] There is yet another reason why capital cannot grow infinitely. If the interest rate r_t fell below zero, consumers would not give all their savings to the firms, but would only supply capital until $r_t = 0$ ($r > 0$ is not possible due to competition between consumers for positive r). So $k \leq \check{k}$ holds always where $\check{k}$ is defined by $r(\check{k}) = \alpha\beta\check{k}^{\alpha-1} - \delta = 0$ (see (8.1)) or $\check{k} = (\alpha\beta/\delta)^{1/(1-\alpha)}$. It follows from (8.14) that $\check{k} > k^*$ if $\alpha/\delta > 1-\alpha$. We assume that δ is sufficiently small to fulfil this condition so that we do not have to bother about $\check{k}$.

where the sign follows from assumption 8.1. As mentioned $X < 0$ is surely satisfied if $1 - w_{k_{t-1}} \geq 0$ or $1 \geq \alpha(1-\alpha)\beta k_{t-1}^{\alpha-1}$ holds. Therefore we derive that X is negative for $k_{t-1} \geq \tilde{k} := [\alpha(1-\alpha)\beta]^{1/(1-\alpha)}$. This is almost identical with $k^* = [(1-\alpha)\beta]^{1/(1-\alpha)}$ except for the additional factor α (see (8.14)). So $\tilde{k}$ is smaller than k^* (see figure 8.6). Therefore for $k_{t-1} \geq \tilde{k}$ assumption 8.1 is a fact, not an assumption. Note that it follows immediately from (8.21) and (8.22) that the $\bar{k}$-line is more negatively sloped than the $a_t^1 = 0$-line at k^* $((dP_t^2/dk_{t-1})|_{a_t^1=0} > (dP_t^2/dk_{t-1})|_{k_t=k_{t-1}}$ since $1 - w_{k_{t-1}} > 0$ at k^*). Hence, starting from k^* the interior $\bar{k}$-line moves north-west (see figure 8.3). At some capital stocks smaller than $\tilde{k}$, however, X might become zero and positive causing the $\bar{k}$-line to move west and south-west. But the $\bar{k}$-line crosses the $a_t^1 = 0$-line only once - at k^* to be precise. Thus the $\bar{k}$-line lies always above the $a_t^1 = 0$-line. Consequently, the $\bar{k}$-line cannot move south-west for very long, but must move north-west again. We simplify by assuming that the $\bar{k}$-line is always moving north-west, i.e. $X < 0$ (assumption 8.1). Below the $\bar{k}$-line P_t^2 is smaller. So according to (8.19) k_t is higher: Capital is growing as indicated by the arrows in figure 8.3. Similarly, above the $\bar{k}$-line capital is falling.

2. On the $\bar{P}_2$-line pollution assimilated by nature is equal to newly generated emissions

$$(1-\theta)P_t^2 = h(k_t + \gamma a_t^1, 0) \tag{8.23}$$

(see (8.3)). For $a_t^1 = 0$ this simplifies to $(1-\theta)P_t^2 = h(k_t, 0) = h(w_{t-1}, 0)$. So we find

$$\left.\frac{dP_t^2}{dk_{t-1}}\right|_{P_{t+1}^2 = P_t^2 \text{ and } a_t^1 = 0} = \frac{h_k w_{k_{t-1}}}{1-\theta} > 0. \tag{8.24}$$

Thus the $\bar{P}^2$-line is rising below the $a_t^1 = 0$-line as shown in figure 8.3. Note that the $\bar{P}^2$-line begins at the origin where it is vertical since $\lim_{k_{t-1}\to 0} w_{k_{t-1}} = \infty$.

For $a_t^1 > 0$ we totally differentiate (8.23) to derive

$$\begin{aligned}(1-\theta)dP_t^2 &= h_k \frac{\partial k_t}{\partial k_{t-1}} dk_{t-1} + h_k \frac{\partial k_t}{\partial P_t^2} dP_t^2 \\ &\quad + \gamma h_k \frac{\partial a_t^1}{\partial k_{t-1}} dk_{t-1} + \gamma h_k \frac{\partial a_t^1}{\partial P_t^2} dP_t^2.\end{aligned}$$

Using (8.19) and (8.20) this yields

$$\left.\frac{dP_t^2}{dk_{t-1}}\right|_{P_{t+1}^2=P_t^2} = \frac{h_k(\partial k_t/\partial k_{t-1} + \gamma \partial a_t^1/\partial k_{t-1})}{(1-\theta) - h_k(1-\gamma)\partial k_t/\partial P_t^2} > 0. \tag{8.25}$$

So the $\bar{P}^2$-line is also rising above the $a_t^1 = 0$-line. The $\bar{P}^2$-line becomes flatter when it crosses the $a_t^1 = 0$-line from below: Compared to (8.24) the denominator of (8.25) is higher and the numerator is smaller (since $\partial k_t/\partial k_{t-1} + \gamma \partial a_t^1/\partial k_{t-1} \leq \partial k_t/\partial k_{t-1} + \partial a_t^1/\partial k_{t-1} = w_{k_{t-1}}$). To the left of the $\bar{P}_2$-line k_{t-1} is smaller and therefore k_t and a_t^1 (if $a_t^1 > 0$) are smaller so that assimilation is higher than new emissions $(1-\theta)P_t^2 > h(k_t + \gamma a_t^1, 0)$. Thus P^2 is falling (see figure 8.3). To the right of the $\bar{P}_2$-line P^2 is rising. □

9. When Abatement Increases Pollution: Far-sighted Governments

9.1 Introduction

In chapter 8 we have looked at the path which is chosen by a succession of myopic governments maximizing only the welfare of the generations living in the same period as the government is in office. Myopic governments neglect four external effects on future generations: First, higher capital and, second, higher abatement of the flow pollutant both increase stock pollution for future generations, two negative externalities, third, higher abatement of the stock pollutant lets stock pollution fall, a positive externality, and, fourth, higher capital increases the wage of the next generation, another positive externality. Far-sighted governments, however, take these external effects into account. Hence, for far-sighted governments abatement of the flow pollutant is less attractive than for myopic governments and abatement of the stock pollutant is more attractive. In fact, far-sighted governments possibly abate emissions of the stock pollutant while myopic ones never do so. Thus far-sighted governments choose a different path than myopic ones. In this chapter we give a few insights into the far-sightedly optimal path and compare it with the myopically optimal path.

Under certain parameter constellations far-sighted governments find that an EKC for the flow pollutant is optimal – as myopic governments do. In contrast, under a completely different parameter constellation – which is not entirely implausible for CO_2 – far-sighted governments choose never to abate the flow pollutant. Instead they abate emissions of the stock pollutant. Therefore flow pollution increases with income: Its PIR is monotonically rising.

Hence, in these cases the myopically optimal EKC is due to the neglect of future damages.

While the fact that abatement of the flow pollutant increases stock pollution (see section 8.1) has almost no influence on the solution of myopic governments, it becomes important for far-sighted governments: The external effect of the abatement technology increases the probability that far-sighted governments abate stock, not flow pollution, i.e. it increases the probability that the PIR for the flow pollutant is monotonically rising for far-sighted governments, but an EKC for myopic ones.

This chapter is organized as follows. In section 9.2 we derive a few characteristics of the far-sightedly optimal path: We show that the PIR for the flow pollutant can be monotonically rising or an EKC depending on parameter constellations. In section 9.3 we compare these two possibilities with the myopically optimal path. In section 9.4 we summarize our findings and we draw some important conclusions based on chapter 3 and part III.

9.2 The solution of far-sighted governments

We analyse the same model as described in section 8.2. An infinitely-lived far-sighted government or a succession of far-sighted governments maximizes

$$\max_{\{a_t^1, a_t^2\}} \sum_{t=1}^{\infty} \rho^{t-1} u[c_t, P_t^1 + \lambda P_t^2]$$

subject to (8.3) – (8.10) where k_0 and P_1^2 are given and again assumed to be small, where u ($\sum \rho^{t-1} u$) is called the felicity function (utility function) as in part II, and where $0 < \rho \leq 1$ is the intergenerational discount rate. This rate can be interpreted as a measure of altruism towards future generations (*Weil* 1987). If ρ was zero, i.e. if there was no altruism, the far-sighted problem would coincide with the myopic problem. Hence, the difference between the models in chapters 8 and 9 is a change in the utility function of the representative consumer which either is altruistic towards future generations or is not. Suppose, however, that in each period the government has a myopic utility function while the representative consumer has a far-sighted one. Then we might suspect that in chapter 8 bad institutions allow the government to maximize its own utility function, but that an improvement in the institutional framework (which occurs outside the model) forces the government to maximize the far-sighted preferences of the representative consumer in chapter 9. According to (8.3) P_t^2 can be written as

$$\begin{aligned} P_t^2 &= \theta P_{t-1}^2 + h(k_{t-1} + \gamma a_{t-1}^1, a_{t-1}^2) \\ &= \theta^2 P_{t-2}^2 + \theta h(k_{t-2} + \gamma a_{t-2}^1, a_{t-2}^2) + h(k_{t-1} + \gamma a_{t-1}^1, a_{t-1}^2) \\ &\vdots \\ &= \theta^{t-1} P_1^2 + \sum_{i=1}^{t-1} \theta^{i-1} h(k_{t-i} + \gamma a_{t-i}^1, a_{t-i}^2). \end{aligned}$$

So the government's maximizing problem becomes

$$\begin{aligned} &\max_{\{k_t, a_t^2\}} \sum_{t=1}^{\infty} \rho^{t-1} \Big\{ u^t \left[c^t(k_t), g^t(k_t, w(k_{t-1}) - k_t - a_t^2) + \lambda \theta^{t-1} P_1^2 \right. \\ &+\lambda \textstyle\sum_{i=1}^{t-1} \theta^{i-1} \, h^{t+1-i}(k_{t-i} + \gamma[w(k_{t-1-i}) - k_{t-i} - a_{t-i}^2], a_{t-i}^2)] \\ &\qquad + \phi_t(w(k_{t-1}) - k_t - a_t^2) + \psi_t a_t^2 \Big\} \end{aligned}$$

where we have introduced some indices (as in u^t) to indicate the period (x is only an exponent in ρ^x and θ^x), where $w(k_{t-1}) = (1-\alpha)\beta k_{t-1}^{\alpha}$ (see (8.2)), and where ϕ_t and ψ_t are the multipliers of the nonnegativity constraints $a_t^1 \geq 0$ and $a_t^2 \geq 0$, respectively. Taking the derivatives with respect to k_t and a_t^2, respectively, the first order conditions for all t read

$$\begin{aligned}
&\underbrace{[u_c^t c_k^t + u_P^t(g_k^t - g_a^t) - \phi_t]}_{=:V} \\
&+ \underbrace{\rho[u_P^{t+1}(g_a^{t+1} w_k^t + \lambda h_k^{t+1}(1-\gamma)) + \phi_{t+1} w_k^t]}_{=:W} \qquad (9.1) \\
&\underbrace{+ \lambda[\theta h_k^{t+1}(1-\gamma) + h_k^{t+2}\gamma w_k^t] \sum_{i=2}^{\infty} \rho^i u_P^{t+i} \theta^{i-2}}_{=:X<0} = 0
\end{aligned}$$

$$\underbrace{[-u_P^t g_a^t - \phi_t + \psi_t]}_{=:Y} + \underbrace{\lambda(h_a^{t+1} - \gamma h_k^{t+1}) \sum_{i=1}^{\infty} \rho^i u_P^{t+i} \theta^{i-1}}_{=:Z>0} = 0. \quad (9.2)$$

These conditions are structured as follows. The term V is the marginal effect of an increase in k_t on immediate felicity, i.e. on u^t: Higher capital increases consumption, but flow pollution rises as well. The term W is the marginal effect of rising k_t on the discounted felicity of the next period, i.e. on ρu^{t+1}: Higher capital allows more a_{t+1}^1 and thus lower flow pollution, but it lets stock pollution grow. The term $X < 0$ is the marginal effect of higher k_t on the discounted felicity of all subsequent periods, i.e. on $\sum_{i=2}^{\infty} \rho^i u^{t+i}$: The higher k_t and the correspondingly higher a_{t+1}^1 increase stock pollution. (9.2) is structured similarly. The term Y is the marginal effect of an increase in a_t^2 on immediate felicity: Flow pollution rises since we abate more stock instead of flow pollution. Finally, the term $Z > 0$ is the marginal effect of growing a_t^2 on the discounted felicity of all subsequent periods, i.e. on $\sum_{i=1}^{\infty} \rho^i u^{t+i}$: Stock pollution declines with higher a_t^2 and also with lower a_t^1 due to the external effect. Without any altruism ρ is zero. Then (9.1) and (9.2) simplify to $V = 0$ and $Y = 0$, respectively. These are the optimality conditions of myopic governments (see (8.12)).

From the first order conditions (9.1) and (9.2) we can immediately draw two conclusions which are stated in the next two propositions.

Proposition 9.1 *There are cases in which far-sighted governments never abate flow pollution so that the PIR for the flow pollutant is monotonically rising.*

Proof. We can write (9.2) as $\psi_t - \phi_t = u_P^t g_a^t - Z$. Thus whenever $u_P^t g_a^t - Z$ is negative, we find $\phi_t > \psi_t \geq 0$. So a_t^1 equals zero. If $a^1 = 0$ holds along the whole path, the PIR for P^1 is monotonically increasing since then $P^1 = g(k, 0)$ depends only on capital. □

Flow pollution is never abated if $u_P^t g_a^t - Z$ is negative, i.e. if the following conditions all hold simultaneously: Abatement of the flow pollutant is expensive ($|g_a|$ small) and has a large external effect on the stock pollutant (γh_k high), abatement of the stock pollutant is cheap ($|h_a|$ high), altruism is high (ρ high), and the stock pollutant is harmful (λ high) and decays slowly (θ high). This is not completely implausible for CO_2 because for this example θ is high as CO_2 has a lifetime of about 125 years (*Frey at al.* 1991, 165), because λ is high since the climate change makes all the problems about air and water quality look tiny (*Beltratti* 1996, 9 and *Nentwig* 1995, 383), and because – as discussed in section 8.1 – end-of-pipe abatement technologies have an external effect on CO_2 (γh_k high).

It is also possible that a^1 is zero and a^2 positive if the abatement technology for the stock pollutant is expensive or even if the stock pollutant cannot be abated at all ($h_a = 0$). Intuitively, this is the case when the government wants to slow down capital growth because the negative external effects of higher capital (higher P^1 in this period and higher P^2 in all subsequent periods) are higher in terms of discounted felicity than the positive effects of higher capital (higher c in this period and higher a^1 in the next period). To slow down growth the government must increase a^1 or a^2 (as $k_t = w(k_{t-1}) - a_t^1 - a_t^2$). However, if the negative long-run effects of a_t^1 (higher P^2 in all subsequent periods) outweigh its positive short-run effect (smaller P^1 in period t), increasing a_t^1 is not advisable. Therefore the government increases a_t^2 even if this has no direct effect on P^2 ($h_a = 0$).

Note that – apart from making the model more realistic – the only reason why we have introduced the external effect of the abatement technology for the flow pollutant on the stock pollutant (i.e. $\gamma > 0$) is that this makes it more probable that far-sighted governments never abate flow pollution.

Under a completely different parameter constellation ($|g_a|$ high and γh_k, $|h_a|$, ρ, λ, θ small), however, we find qualitatively the same result as for myopic governments:

Proposition 9.2 *There are cases in which stock pollution is never abated. Then far-sighted governments choose an EKC for the flow pollutant and a monotonically increasing PIR for the stock pollutant.*

Proof. If $u_P^t g_a^t - Z$ is positive along the whole path, $\psi_t > \phi_t \geq 0$ holds such that a_t^2 is zero (see (9.2)). Note that a^1 decreases not only P^1, but also P^2 since more a^1 means less k such that $h(k + \gamma a^1, 0)$ is smaller as $\gamma \leq 1$. Hence, if a^2 is always zero, the behaviour of the economy is similar to the myopically optimal path (cf. proof of proposition 8.1): First, there is no abatement and the economy grows.[1] Then a^1 becomes positive. When the economy reaches the steady state capital stock, P^2 did not yet have the time to accumulate and is therefore smaller than in the steady state. Thus capital overshoots. When the maximal capital stock is reached, only P^2 increases causing a^1 to rise as well. So we again find an EKC for P^1 in this case. The optimal path looks as in figure 8.5. □

Until now we have only considered the two extreme cases in which either a^1 or a^2 remains zero. However, these two cases already reveal the main result: For far-sighted governments the PIR for the flow pollutant might be an EKC or monotonically

[1] Assuming $\lim_{c\to 0} u_c = \infty$ and dividing through (9.1) by w_k^t it follows from $\lim_{k\to 0} w_k = \infty$, from $\lim_{k\to 0} c_k/w_k = \alpha/(1-\alpha)$ (see (8.2) and (8.11)), and from the fact that all other terms in (9.1) – excluding ϕ – are bounded that $\phi_t > 0$ and thus $a_t^1 = 0$ for small k.

rising. Hence, it is not worth deriving any further details of the far-sighted solution. Instead we compare these two extremes with the myopically optimal path in the next section.

9.3 A comparison of the far-sighted and the myopic solution

When far-sighted governments do not abate flow pollution (proposition 9.1), the PIR for the flow pollutant is monotonically rising. In contrast, an EKC emerges if myopic governments are in office. The result of myopic governments – an EKC for the flow pollutant and a rising PIR for the stock pollutant – is compatible with the empirical evidence. This, however, might not be optimal from a long-run point of view. Hence, *observed EKCs might be due to the neglect of future damages.*

However, proposition 9.2 shows that the EKC for the flow pollutant can also be optimal from a far-sighted point of view. Still, there are two differences between the far-sightedly and myopically optimal path when stock pollution is never abated on both paths:

Proposition 9.3 *Suppose that a^2 is always zero and that $W + X > 0$ always holds, i.e. that more capital in this period has a positive marginal effect on the discounted felicity of all subsequent periods. Then starting with the same capital and pollution stock far-sighted governments start to increase a^1 later on than myopic governments. Furthermore, the steady state of far-sighted governments lies at higher levels of capital and of P^2 than the steady state of myopic governments.*

Proof. Myopic governments for which ρ is zero choose capital such that $V = 0$ holds (see (9.1)). For far-sighted governments, however, (9.1) is positive at this capital stock if $W+X > 0$: Increasing k_t makes utility higher. Therefore far-sighted governments always choose a higher capital stock and accordingly a smaller a^1 than myopic governments. Two conclusions emerge. First, far-sighted

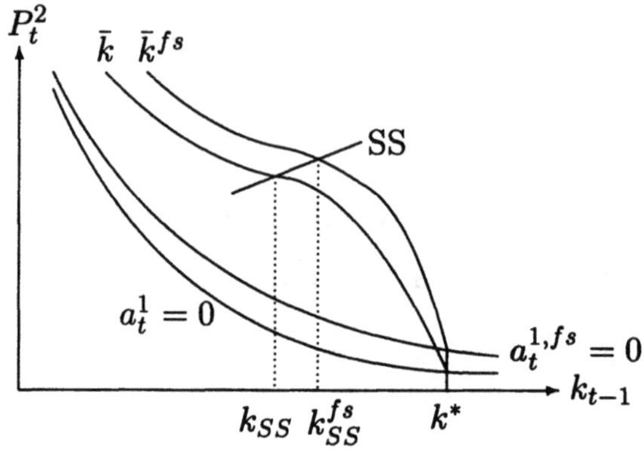

Figure 9.1: Comparison of the myopic and far-sighted solution if $a^2 = 0$ and $W + X > 0$.

governments also prefer higher capital when in the myopic solution a^1 just becomes positive, i.e. when the optimal path in figure 8.5 crosses the $a_t^1 = 0$-line. Thus far-sighted governments start to abate later on: The $a_t^{1,fs} = 0$-line lies above the $a_t^1 = 0$-line in figure 9.1 (where here and later in the proof fs stays for far-sighted).

Second, suppose that the myopic government keeps capital constant (corresponding to a point on the $\bar{k}$-line in figure 8.3 or 9.1). Then the far-sighted government also chooses higher capital. So capital grows. If P_t^2 was higher, the myopic government would let capital fall, but the far-sighted government would again pick a higher capital stock. Thus the $\bar{k}^{fs}$-line lies above the $\bar{k}$-line as shown in figure 9.1.

At the steady state (SS) of myopic and of far-sighted governments pollution assimilated by nature must equal newly generated emissions, i.e. $(1-\theta)P_{SS}^2 = h(k_{SS} + \gamma(w(k_{SS}) - k_{SS}), 0)$ which follows from (8.3). Since $dP_{SS}^2/dk_{SS} = h_k(1 - \gamma + \gamma w_k)/(1-\theta)$ is positive, both steady states must lie on an upward sloping line called the SS-line in figure 9.1. Both steady states lie at the intersection of the SS-line with the $\bar{k}$- or $\bar{k}^{fs}$-line. Hence, since the

$\bar{k}^{fs}$-line lies above the $\bar{k}$-line, the steady state of the far-sighted government lies at higher levels of capital and of P^2 (see figure 9.1). □

If P^1 starts to decline as soon as a^1 becomes positive on both the myopically and far-sightedly optimal path, the pollution and income level at the turning point of the EKC for P^1 is higher for far-sighted governments if $W + X > 0$. The only positive term in $W + X$ is $\rho u_P^{t+1} g_a^{t+1} w_k^t$ (see (9.1)). Thus $W + X$ is positive, if this term is large relative to the other terms, i.e. if the following conditions all hold simultaneously: Abating flow pollution is cheap ($|g_a|$ high), capital and abatement of the flow pollutant are only slightly polluting (h_k small), altruism is low (ρ small), flow pollution is harmful relative to stock pollution (λ small), and stock pollution decays fast (θ small).[2] As shown above a^2 is zero under similar conditions. So if a^2 equals zero, it is likely that $W+X>0$ holds as assumed in proposition 9.3.

As mentioned, however, λ and θ are high for CO_2. Then a^1, not a^2 tends to be zero and $W+X$ tends to be negative. Hence, in this case far-sighted governments start to abate earlier on, but they abate stock, not flow pollution. In doing so they slow down capital growth because capital has a negative effect on future generations (as $W + X < 0$).

The solutions of myopic and of far-sighted governments are also strongly different if there are binding thresholds. When a threshold for a pollutant is overstepped, large irreversible damages occur like the extinction of animal and plant species or the flip of a whole ecosystem. While a threshold for P^1 is not exceeded by both governments since the damage occurs immediately,[3] a threshold for P^2 is not taken into account by myopic governments:

[2]For $\lambda = 0$ or for $h_k = 0$ all negative terms in $W + X$ disappear and there is no stock pollution.

[3]This is not necessarily true if the irreversible damage is only small. Myopic governments follow an EKC: In the steady state flow pollution can be below the threshold level. The turning point, however, might lie above the threshold level since the (immediate, but irreversible) damage is small. Then myopic

The damage has either already occurred or will only materialize in the future. Therefore myopic governments overstep a binding threshold and end up in a far worse steady state than far-sighted governments that make sure that the threshold is not exceeded. However, it might be argued that the very existence of a threshold makes far-sightedness politically feasible.

9.4 Conclusion

Contrary to myopic governments who never abate stock pollution far-sighted governments find that under certain parameter constellations it is preferable to abate stock, not flow pollution. This is the case if abatement of the flow pollutant is expensive and has a large external effect on the stock pollutant, if abatement of the stock pollutant is cheap, if altruism towards future generations is high, and if the stock pollutant is harmful and decays slowly. This is not completely implausible for CO_2 which is harmful, decays slowly, and is emitted by end-of-pipe abatement technologies for flow pollutants (as shown in section 8.1). If far-sighted governments abate only the stock pollutant, the PIR for the flow pollutant is monotonically rising. Thus in this case the EKC obtained by myopic governments is caused by the neglect of future damages.

Under a completely different parameter constellation, in contrast, far-sighted governments also never abate the stock pollutant. Then the PIR for the flow pollutant is again an EKC. Under this parameter constellation it is likely that far-sighted governments start to abate flow pollution later on than myopic governments. The underlying reason is that far-sighted governments take into account that higher capital allows for higher abatement expenditures in the next period. This effect is larger than the effect of internalizing the external effects on the stock pollutant under this parameter constellation. In contrast, under a different param-

governments let the damage occur, but far-sighted ones keep flow pollution below the threshold.

eter constellation far-sighted governments start to abate earlier on than myopic governments. Then, however, far-sighted governments tend to abate stock, not flow pollution.

As mentioned in chapter 1 the finding of an EKC for certain pollutants has often been seen as a sign for optimism. Thus 'most governments and global institutions see no conflict between economic growth [and environmental degradation]' (*Cole* 1999, 91). However, in part III we have shown that the EKC for a flow pollutant might be due to the neglect of future damages and due to ever rising stock pollution. Therefore we claim that great care must be taken when interpreting the results of EKC studies. In particular, we cannot conclude from the finding of an EKC for some (flow) pollutants that other (stock) pollutants also follow an inverted-U shaped path. Furthermore, falling flow pollution might only be achieved at (high) long-run costs. Hence, if the EKC for flow pollution is actually caused by the neglect of future damages and by ever rising stock pollution, the EKC is bad news for future generations. Indeed, in this case, since all observed EKCs are based on past data, the EKC is bad news for us.

We can also draw some important conclusions based on chapter 3 and part III of this book. The main conclusion of chapter 3 is that for a certain pollutant an EKC only exists when policy measures are taken with respect to this pollutant, but that the PIR is monotonically rising when the government remains inactive with regard to this pollutant. In section 8.1 we have argued that it is impossible to internalize all external effects because of coordination problems and transaction costs and because some damages cannot be easily measured. Similarly, *Huesemann* (2001, 274) argues that 'considering the limited budgets for environmental research, it is obvious that only a few selected [environmental problems] can ever be studied'. Therefore *Huesemann* (2001, 278) concludes that 'in practice [...] only the obvious problems are addressed, i.e. those situations where sufficient scientific evidence is available to indicate that environmental pollutants [...] might pose a risk to humans or whatever human values. All other environmental problems [...] are ignored'.

Hence, on the one hand, policy measures and accordingly an EKC can only be expected for a few selected pollutants. These tend to be those (flow and local) pollutants which currently cause high damages. For these pollutants policy measures find a majority in the electorate and we observe an EKC. However, as mentioned in section 3.11 policy measures tend to be delayed responses to (severe) environmental damages because the government responds to public concern which only arises when damages are (clearly) visible and when sufficient scientific knowledge is available to find out which substance (or which mixture of substances) is responsible for the damage.

On the other hand, there are always unregulated (stock and global) pollutants and unregulated potential pollutants, i.e. unregulated substances whose effects on the environment are not yet known. Firms have no incentives to abate these pollutants such that their PIR is expected to be monotonically rising. Due to the internalization of a (flow or local) pollutant causing high damages, firms may even substitute this pollutant by another, unregulated (stock, global, or potential) pollutant – possibly because they concentrate all their thoughts and energy to tackle the regulated pollutant, i.e. possibly because they are not aware that they emit more of an unregulated pollutant. Later on, when this pollutant also results in high damages (for example, CFCs and CO_2, see subsection 3.10.2), the government will tackle it and the firms will tend to emit yet another unregulated pollutant. Hence, there are always pollutants exhibiting a monotonically rising PIR. It follows that *the EKC is not valid for all pollutants.* The empirical evidence presented in paragraph 2.3.2.1 also shows that *the EKC is not valid for aggregate pollution. Huesemann* (2001, 274) even states that 'the focus of current environmental science is too narrowly anthropocentric to truly ensure the long-term protection of human health and welfare.'

Furthermore, when we try to estimate the PIR we tend to find only data for those pollutants which cause high damages because for these pollutants data collection is worth its costs. For these pollutants the PIR is likely to be an EKC. For unregulated (stock, global, and potential) pollutants, however, data are often not available. Then the PIR which presumably is upward sloping cannot be estimated due to a lack of data. Hence, *empirical studies tend to paint a too optimistic image.* This is corroborated by the analysis of some aggregate measures of pollution in paragraph 2.3.2.1 whose PIRs are monotonically rising.

10. Conclusion

In the last decade many studies have analysed the relationship between pollution and income. It has been found that the pollution-income relationship (PIR) is monotonically rising for some pollutants, but that the PIR is inverted-U shaped for other pollutants. This inverted-U shaped PIR is called an environmental Kuznets curve (EKC). The main purpose of this book has been to identify possible causes of the EKC. In other words, our aim has been to find out why the PIR is an EKC for some pollutants, but monotonically increasing for other pollutants.

Before addressing this question, however, we have surveyed the empirical literature on the EKC in chapter 2. We have shown that an EKC is observed for air pollutants such as sulphur dioxide (SO_2), suspended particulate matter (SPM), oxides of nitrogen (NO_x), carbon monoxide (CO), lead, volatile organic compounds (VOC), and – although the evidence is less clear – for some river water pollutants. These are all short-lived pollutants, i.e. flow pollutants. Moreover, they cause mainly local damages. However, the PIR is monotonically rising for carbon dioxide (CO_2) and for waste which are both long-lived, i.e. stock pollutants. CO_2 is also a global pollutant. Furthermore, for aggregate measures of pollution the PIR is monotonically increasing as well. Some studies have also obtained N-shaped PIRs: The PIR first rises, then falls, and finally rises again. The final upturn means that in the end income growth is bad for the environment.

We have also critically examined the estimation technique with which these results have been derived. One of the main problems is finding reliable and comparable data for pollution. In fact, data are missing for many dimensions of environmental quality such as soil erosion, desertification, pollution and depletion of groundwater aquifers, biodiversity loss, acidification, extinction of animal and plant species, climate change, the ozone hole, the risk of nuclear power stations, and ecosystem resilience, which is a measure of the magnitude of shocks an ecosystem can accommodate without flipping into another, possibly environmentally disastrous equilibrium. From an econometric point of view we have shown that the use of a time trend can be problematic (we might think that we found an EKC although the true PIR is N-shaped), that multicollinearity might hamper the results (slight variations in the (unreliable) data can have dramatic effects on the results), and that unit roots might pose a problem if there is no cointegration (then the results are not valid). Furthermore, most studies assume that the turning point of the EKC occurs at the same income level in all countries. However, recent studies show that this assumption is wrong: Every country has its own PIR.

In chapter 3 we have reviewed the theoretical literature looking for possible causes of the EKC. We have also assessed the empirical validity of these causes. Economic growth leads to more output and thus to higher pollution *ceteris paribus.* This so-called scale effect can, however, be counteracted by several forces. There is empirical evidence that demand for environmental quality increases with income. Provided that relatively advanced political institutions exist, higher demand for environmental quality causes ever stricter environmental regulations. The evidence suggests that the income distribution is not crucial for the strictness of policy measures. However, a more equal distribution of political power might help causing the downturn of the EKC. The internalization of external effects might be a continuous process or it might occur in sudden jumps (shocks).

Internalization can cause technological progress to move in an environmentally friendly direction. Empirically, changing technologies have been crucial for the downturn of the EKC. However, when appropriate policies are missing, technological progress might well be pollution-enhancing. The introduction of a new technology due to the internalization of the external effects of a certain (flow or local) pollutant might also result in higher emissions of another unregulated (stock or global) pollutant. Then one pollutant is only substituted by another. In fact, end-of-pipe technologies have exactly this effect because they use energy and thus increase CO_2 emissions. A possible explanation of the EKC is also that increasing returns to scale in abatement might allow rich countries to abate at lower average costs than poor countries.

There is evidence that dirty industries migrate from developed into developing countries. Thus rich countries can reduce pollution by relocating production and by importing the goods. If the cross country EKC is based on migration, the EKC shifts upwards and to the right when all countries grow simultaneously. It has also been argued that structural change – the rise and fall of the heavy industry – can explain the EKC. However, the empirical evidence does not support this hypothesis: Dirty industries are not shrinking in developed countries. It follows that outmigration of dirty industries is not fast. Hence, the migration of dirty industries might explain the EKC estimated in a cross country analysis, but it cannot explain the EKC of a single developed country over time.

The literature survey has therefore revealed four empirically validated causes of the EKC of a single country over time: Increasing demand for environmental quality, technological progress, substitution between pollutants, and increasing returns to scale in abatement. Satisfying the increasing demand for environmental quality requires policy measures, technological progress is only environmentally friendly if appropriate incentives are set by policy makers, substitution is caused by regulations or technological progress, and increasing returns to scale in abatement are only important when there is abatement which is possibly not the case without policy measures. Hence, we conclude that environmental policy is of utmost importance for the downturn of the EKC: *For a certain pollutant an EKC only exists when policy measures are taken but the PIR is ever rising if the government remains inactive.* Hence, the idea that economic growth is good for the environment is not tenable. *Ayres* (1995, 97) even claims that this idea is 'false and pernicious nonsense'. It also follows that fighting corruption might be important since corruption might postpone the implementation of environmental policies.

However, in most cases emission reductions are due to *local* environmental policy reforms which often ignore international and intergenerational consequences. Hence, we find an EKC for flow and local pollutants, but a monotonically rising PIR for stock and global pollutants. Furthermore, policy measures tend to be taken as a belated response to evidence of (severe) environmental damages. This is possibly the case because policy makers respond to public concern which arises when scientific knowledge about the sources of these damages becomes available and when the damages are (clearly) visible and because the process of finding acceptable regulations is time-consuming. Nevertheless, it has been argued that policy makers should prevent pollution from rising to high pollution levels. This is important because due to irreversibilities even the falling branch of the EKC does not necessarily imply that environmental quality is improving with rising income: Once an irreversible damage has occurred, even zero pollution is not sufficient to restore the pristine state of the environment.

Before continuing the analysis of the EKC we have studied two methodological issues in chapter 4. First, we have developed a method how we can calculate the results of a static model with only ordinal, but no cardinal assumptions about the utility function. Many researchers make stronger, cardinal assumptions. Our method prevents interpretations which violate the normal good assumption, i.e. the assumption that demand for all goods increases with income. The method also allows to simplify equations. Second, we have derived a method how we can figure out which results of a dynamic model are robust to changes in the elasticity of intertemporal substitution. If the elasticity of intertemporal substitution is small, we prefer a smooth development path. In chapters 5–8 we have employed these two methods.

In chapters 5–7 we have looked more closely at one of the empirically validated causes of the EKC: Rising demand for environmental quality. It is often argued in the literature that an EKC emerges because environmental quality is a luxury good. In a theoretical model we have shown in chapter 5 that what really matters is not that the income elasticity of demand for environmental quality is greater than one (i.e. that environmental quality is a luxury good), but that it is greater than zero (i.e. that environmental quality is a normal good). As we have argued the normality of environmental quality can also be called a tendency to satiation in consumption, i.e. a tendency to zero marginal utility of consumption at a high, possibly infinite consumption level.

To derive this result we have used the simple static model of *McConnell* (1997) and we have relaxed several of his assumptions. Thus we have analysed a representative consumer model with pollution generated by consumption. We have found that at low income levels no abatement is optimal and pollution increases with income. Once abatement expenditures are positive, we have demonstrated that satiation in consumption is not only sufficient to find an EKC, but a tendency to satiation (i.e. the normality of environmental quality) is even *necessary* if we assume a standard functional form for the pollution function. The intuition is that approaching the satiation level we can easily dispose of some consumption almost without lowering utility. Therefore additional income is better used for abatement such that pollution falls. We have also proven that the model of *Stokey* (1998) is a special case of our model. A new interpretation of *Stokey's* (1998) result has been offered: With asymptotic satiation there is an EKC, but without satiation pollution is monotonically rising with income. Thus with her specification of functional forms the shape of the PIR exactly depends on whether or not there is satiation.

Because the EKC is an inherently dynamic phenomenon in the course of economic growth, we have made the model dynamic in chapter 6. The standard Ramsey approach has been used to endogenize growth. Surprisingly, almost all results of the static model remain valid in the dynamic framework. The steady state pollution level is zero if the following conditions all hold simultaneously: The steady state income level is high due to a highly productive technology, there is a tendency to satiation, and abatement is cheap. We have also derived that almost all results are robust to changes in the intertemporal elasticity of substitution. In particular, the shape of the PIR (i.e. whether it is monotonically rising, an EKC, or N-shaped), the pollution level at the turning point of the EKC (if there is one), and the steady state are all robust. Only the income level at the turning point of the EKC rises with the elasticity of intertemporal substitution.

However, the elasticity of intertemporal substitution plays a crucial role for the PIR in the model of chapter 7. To arrive at this result we have considered the same model as in chapter 6, but we have assumed that pollution is generated by production capital (instead of consumption). Thus we have examined an extension of the model of *Selden* and *Song* (1995). These authors conclude that an EKC emerges from their model. This conclusion, however, is only correct when the elasticity of intertemporal substitution is sufficiently high. If the elasticity of intertemporal substitution is lowered, the PIR becomes monotonically rising or N-shaped. Intuitively, when the elasticity of intertemporal substitution is small, a smooth development path is preferred. Therefore we let pollution develop smoothly, possibly monotonically, and we do not accept high pollution levels along the path which would result in an EKC. Hence, if pollution is generated by production capital, the shape of the PIR (i.e. whether it is monotonically rising, an EKC, or N-shaped) can change with the elasticity of intertemporal substitution.

Furthermore, we have shown that the steady state is robust to changes in the elasticity of intertemporal substitution and that the steady state pollution level is zero under almost the same conditions as in chapter 6. If the steady state pollution level is zero, the PIR is an EKC. However, the level of pollution and of income at the turning point of the EKC both rise with the elasticity of intertemporal substitution.

In chapters 8 and 9, finally, we have tried to find an explanation for the empirical evidence that there is an EKC for flow pollutants, but that the PIR for stock pollutants is monotonically rising. We have found a new possible cause of the EKC: The EKC for a flow pollutant might be due to the neglect of future damages and due to ever rising stock pollution. To derive this result we have extended the overlapping generations model of *John* and *Pecchenino* (1994) to two pollutants: Whereas the flow pollutant causes immediate damages, the stock pollutant harms the environment only in the future. In chapter 8 we have analysed the solution of a succession of myopic governments and in chapter 9 we have sketched the far-sighted optimum.

Myopic governments never abate the stock pollutant. Thus on the myopically optimal path the PIR for the stock pollutant is monotonically rising. Therefore to keep aggregate pollution at reasonably low levels myopic governments abate flow pollution. So the PIR of the flow pollutant is an EKC. Without the stock pollutant, in contrast, the PIR of the flow pollutant can be monotonically rising: Rising stock pollution can cause the downturn of the EKC for the flow pollutant. Moreover, if stock pollution is high and harmful, the turning point of the EKC for the flow pollutant lies at lower levels of income and of flow pollution. Since the level and the harmfulness of stock pollution differ between countries, the model casts doubts on most empirical EKC studies because they assume that the turning point occurs at the same income level in all countries. However, it is consistent with recent empirical studies which obtain that the PIR differs between countries.

As shown in chapter 9 far-sighted governments might also find that an EKC for the flow pollutant is optimal. However, it is also possible that far-sighted governments prefer to abate stock pollution and thus never abate the flow pollutant. Then the PIR of the flow pollutant is monotonically rising. Hence, in this case the myopically optimal EKC is due to the neglect of future damages.

Finally, we have argued that the myopic solution – which is consistent with the empirical evidence – might be closer to reality than the far-sighted solution. There are good reasons to believe so: It is impossible to internalize all environmental damages, first, because of coordination problems and transaction costs which make it simply infeasible to closely monitor all (potential) pollutants and, second, because some damages are difficult to measure. So there are always unregulated pollutants. These pollutants tend to be stock or global pollutants since the incentives to monitor them are smaller than for flow or local pollutants. In chapter 3 we have argued that for a certain pollutant an EKC only arises when policy measures are taken, but that the PIR is monotonically rising if the government remains inactive. Hence, an EKC can only be expected for those pollutants which currently cause high damages because for these pollutants policy measures tend to be implemented. For all other pollutants the PIR tends to be monotonically rising. In fact, the empirical evidence shows that the PIR for most aggregate measures of pollution is monotonically increasing.

Summarizing our results we have found several empirically validated causes of the EKC. The evidence suggests that the migration of dirty industries can only explain an EKC obtained in a cross country analysis. For an EKC in a single country over time we have found four possible causes: First, demand for environmental quality rises with income. A tendency to satiation in consumption is necessary for demand for environmental quality to rise with income. Furthermore, an EKC might only exist if the elasticity of intertemporal substitution is sufficiently high. Second, the EKC can be caused by the substitution of pollutants. In fact, an EKC for a flow pollutant might only exist because stock pollution is rising and because future damages are neglected. Third, the EKC can be explained by technological progress. Fourth, increasing returns to scale in abatement can be responsible for the downturn of the EKC. We have seen that all these causes can only explain the EKC when appropriate policy measures are taken. Regulations, however, often neglect international and intergenerational consequences. Therefore we tend to find an EKC for local or flow pollutants only. For global or stock pollutants, in contrast, we expect a monotonically rising PIR.

Appendix A

Two examples for the pollution function

Throughout part II and part III of this book we use a pollution function of the form $P(C, A)$ or $P(K, A)$: Pollution P is generated by consumption C (in chapters 5 and 6) or capital K (in chapters 7-9) and can be abated by abatement expenditures A. In this appendix we analyse two examples for such a pollution function. We examine $P(C, A)$, but for $P(K, A)$ we can just replace C by K in all equations below. The examples are two standard functional forms which satisfy our assumptions $P_C > 0$, $P_A < 0$, $P_{CC} \geq 0$, $P_{AA} > 0$, and $P_{AC} \leq 0$. We show that the two examples fulfil conditions (5.1) and (5.2). Thus P_C and $|P_A|$ fall when we move along the iso-pollution lines to the right (see figure 5.1 on page 116): Consumption becomes less polluting and abatement becomes more expensive. We also prove that for the first example (6.3) and (7.2) are always satisfied whereas for the second example they hold for a vast amount, but not all parameter constellations. Thus in most cases the iso-pollution lines and the pollution function are both convex.

The **first example** is taken from *Stokey* (1998). Pollution is given by

$$P = C^{\beta}(C + A)^{1-\beta} - x$$

where $\beta > 1$ and $x \geq 0$. For a derivation of this function see section 5.4. Denoting $C + A$ by income Y we find (note that $A \geq 0$ and therefore $Y \geq C$ must hold)

$$
\begin{aligned}
P_C &= \beta C^{\beta-1}Y^{1-\beta} - (\beta-1)C^{\beta}Y^{-\beta} \\
&= C^{\beta-1}Y^{-\beta}(\beta Y - (\beta-1)C) > 0 && \text{(A.1)} \\
P_A &= -(\beta-1)C^{\beta}Y^{-\beta} < 0 && \text{(A.2)} \\
P_{CC} &= \beta(\beta-1)(C^{\beta-2}Y^{1-\beta} - 2C^{\beta-1}Y^{-\beta} + C^{\beta}Y^{-\beta-1}) \\
&= \beta(\beta-1)C^{\beta-2}Y^{-\beta-1}(Y-C)^2 \geq 0 \\
P_{AA} &= \beta(\beta-1)C^{\beta}Y^{-\beta-1} > 0 \\
P_{AC} &= \beta(\beta-1)(C^{\beta}Y^{-\beta-1} - C^{\beta-1}Y^{-\beta}) \\
&= \beta(\beta-1)C^{\beta-1}Y^{-\beta-1}(C-Y) \leq 0
\end{aligned}
$$

with the assumed signs ($P_{CC} = 0$ and $P_{AC} = 0$ hold only for $A = 0$). Note also that $\lim_{A\to 0} |P_A| = \beta - 1 < \infty$. Both P_C and $|P_A|$ fall when we move along the iso-pollution lines to the right because the numerator in (5.1) simplifies to

$$
\begin{aligned}
-P_{CC}P_A + P_{AC}P_C &= \beta(\beta-1)^2(C^{2\beta-2}Y^{1-2\beta} - 2C^{2\beta-1}Y^{-2\beta} \\
&\quad +C^{2\beta}Y^{-2\beta-1}) + \beta^2(\beta-1)(C^{2\beta-1}Y^{-2\beta} - C^{2\beta-2}Y^{1-2\beta}) \\
&\quad -\beta(\beta-1)^2(C^{2\beta}Y^{-2\beta-1} - C^{2\beta-1}Y^{-2\beta}) \\
&= (C^{2\beta-1}Y^{-2\beta} - C^{2\beta-2}Y^{1-2\beta})(\beta^2(\beta-1) - \beta(\beta-1)^2) \\
&= C^{2\beta-2}Y^{-2\beta}(C-Y)\beta(\beta-1) < 0 \qquad \text{for } A > 0 && \text{(A.3)}
\end{aligned}
$$

and the numerator in (5.2), i.e. $-P_{AC}P_A + P_{AA}P_C$, reduces to

$$
\begin{aligned}
&\beta(\beta-1)^2(C^{2\beta}Y^{-2\beta-1} - C^{2\beta-1}Y^{-2\beta}) \\
&+\beta^2(\beta-1)C^{2\beta-1}Y^{-2\beta} - \beta(\beta-1)^2C^{2\beta}Y^{-2\beta-1} \\
= \;& C^{2\beta-1}Y^{-2\beta}(\beta^2(\beta-1) - \beta(\beta-1)^2) \\
= \;& C^{2\beta-1}Y^{-2\beta}\beta(\beta-1) > 0. && \text{(A.4)}
\end{aligned}
$$

As $P_C/P_A = -\beta/(\beta-1)YC^{-1}+1$ and $P_C^2/P_A^2 = \beta^2/(\beta-1)^2Y^2C^{-2}$ $-2\beta/(\beta-1)YC^{-1} + 1$, the numerator of (6.3) is

$$
\begin{aligned}
&P_{CC} - 2P_{AC}P_C/P_A + P_{AA}P_C^2/P_A^2 = \beta(\beta-1)(C^{\beta-2}Y^{1-\beta} \\
&\quad -2C^{\beta-1}Y^{-\beta} + C^{\beta}Y^{-\beta-1}) + 2\beta^2(C^{\beta-1}Y^{-\beta} - C^{\beta-2}Y^{1-\beta}) \\
&\quad -2\beta(\beta-1)(C^{\beta}Y^{-\beta-1} - C^{\beta-1}Y^{-\beta}) \\
&\quad +\beta^3/(\beta-1)C^{\beta-2}Y^{1-\beta} - 2\beta^2C^{\beta-1}Y^{-\beta} + \beta(\beta-1)C^{\beta}Y^{-\beta-1} \\
&= C^{\beta-2}Y^{1-\beta}\left(\beta^3/(\beta-1) + \beta(\beta-1) - 2\beta^2\right) \\
&= C^{\beta-2}Y^{1-\beta}\beta/(\beta-1)\left(\beta^2 + (\beta-1)^2 - 2\beta(\beta-1)\right) \\
&= C^{\beta-2}Y^{1-\beta}\beta/(\beta-1)\,(\beta-(\beta-1))^2 \\
&= C^{\beta-2}Y^{1-\beta}\beta/(\beta-1) > 0.
\end{aligned}
$$

So the iso-pollution lines are convex. Even $P(C, A)$ in (7.2) is marginally convex as

$$
\begin{aligned}
P_{AA}P_{CC} - P_{AC}^2 &= \beta^2(\beta-1)^2\left[(C^{2\beta-2}Y^{-2\beta} - 2C^{2\beta-1}Y^{-2\beta-1}\right. \\
&\quad +C^{2\beta}Y^{-2\beta-2}) - (C^{2\beta}Y^{-2\beta-2} \\
&\quad \left.-2C^{2\beta-1}Y^{-2\beta-1} + C^{2\beta-2}Y^{-2\beta})\right] = 0.
\end{aligned}
$$

The **second example** is

$$
P = \frac{fC^g}{(kA+1)^h} - x,
$$

where $g \geq 1$, $x \geq 0$ and $f, h, k > 0$. Again, $-x$ is included to make zero pollution technically feasible. For $A = 0$ this simplifies to $P = fC^g - x$. We obtain

$$
\begin{aligned}
P_C &= gfC^{g-1}/(kA+1)^h > 0 \\
P_A &= -hkfC^g/(kA+1)^{h+1} < 0 \\
P_{CC} &= g(g-1)fC^{g-2}/(kA+1)^h \geq 0 \\
P_{AA} &= h(h+1)k^2fC^g/(kA+1)^{h+2} > 0 \\
P_{AC} &= -ghkfC^{g-1}/(kA+1)^{h+1} < 0 \\
\lim_{A\to 0}|P_A| &= hkfC^g < \infty
\end{aligned}
$$

in accordance with our assumptions. In this example as well P_C and $|P_A|$ fall when we move along the iso-pollution lines to the right because (see (5.1) and (5.2))

$$\begin{aligned} -P_{CC}P_A + P_{AC}P_C &= \frac{(g^2-g)hkf^2C^{2g-2}}{(kA+1)^{2h+1}} - \frac{g^2hkf^2C^{2g-2}}{(kA+1)^{2h+1}} \\ &= \frac{-ghkf^2C^{2g-2}}{(kA+1)^{2h+1}} < 0 \\ -P_{AC}P_A + P_{AA}P_C &= -\frac{gh^2k^2f^2C^{2g-1}}{(kA+1)^{2h+2}} + \frac{g(h^2+h)k^2f^2C^{2g-1}}{(kA+1)^{2h+2}} \\ &= \frac{ghk^2f^2C^{2g-1}}{(kA+1)^{2h+2}} > 0. \end{aligned}$$

Since P_C/P_A equals $-g(kA+1)/(hkC)$, the numerator of (6.3) simplifies to

$$P_{CC} - 2P_{AC}P_C/P_A + P_{AA}P_C^2/P_A^2 = \frac{(g^2-g)fC^{g-2}}{(kA+1)^h} - \frac{2g^2fC^{g-2}}{(kA+1)^h} + \frac{g^2(h+1)fC^{g-2}}{h(kA+1)^h} = \frac{g(g-h)fC^{g-2}}{h(kA+1)^h}$$

which is positive if $g > h$, but negative if $h > g$. Therefore the iso-pollution lines are convex if $g > h$. However, the iso-pollution lines are concave if the abatement technology is efficient, i.e. if $h > g \geq 1$. For $g = h$ the iso-pollution lines are straight lines. In this example it is easy to show that convexity of $P(C, A)$ is a stronger condition than convexity of the iso-pollution lines because $P_{AA}P_{CC} - P_{AC}^2$ in (7.2) is positive if

$$\frac{(g^2-g)(h^2+h)k^2f^2C^{2g-2}}{(kA+1)^{2h+2}} - \frac{g^2h^2k^2f^2C^{2g-2}}{(kA+1)^{2h+2}} = \frac{gh(g-h-1)k^2f^2C^{2g-2}}{(kA+1)^{2h+2}} > 0$$

or if $g > h+1$. This is a stronger condition than $g > h$ above. So for $h+1 > g > h$ the iso-pollution lines are convex, but $P(C, A)$ is not. Nevertheless, (6.3) and (7.2) are satisfied for $g \geq h+1$, i.e. for many parameter constellations.

References

Agras, Jean and Duane Chapman (1999): „A dynamic approach to the Environmental Kuznets Curve hypothesis". *Ecological Economics* 28, 267–277.

Ahmed, Iftikhar (1995): „Sustainable Livelihood and Employment: Pragmatic Approaches". In: Iftikhar Ahmed and Jacobus A. Doeleman (eds.): *„Beyond Rio: The Environmental Crisis and Sustainable Livelihoods in the Third World"*. ILO Study Series, New York: St.Martin's Press, 317–362.

Andreoni, James and Arik Levinson (2001): „The simple analytics of the environmental Kuznets curve". *Journal of Public Economics* 80, 269–286.

Ang, B.W. (1994): „Decomposition of industrial energy consumption: The energy intensity approach". *Energy Economics* 16 (3), 163–174.

Ang, B.W. and S.Y. Lee (1994): „Decomposition of industrial energy consumption: Some methodological and application issues". *Energy Economics* 16 (2), 83–92.

Ansuategi, Alberto (2000): *„Economic Growth and Environmental Quality: A Critical Assessment of the Environmental Kuznets Curve Hypothesis"*. Thesis, University of York, Environment Department.

Ansuategi, Alberto, Edward Barbier, and Charles Perrings (1998): „The environmental Kuznets curve". In: J. C. J. M. van den Berg and M. W. Hofkes (eds.): *„Economic Modelling of Sustainable Development: Between Theory and Practice"*, Dordrecht: Kluwer Academic Publisher, 139–164.

Ansuategi, Alberto and Charles Perrings (2000): „Transboundary Externalities in the Environmental Transition Hypothesis". *Environmental and Resource Economics* 17 (4), 353–373.

Ansuategi, Alberto and Charles Perrings (2000): „Transboundary Externalities in the Environmental Transition Hypothesis". *Environmental and Resource Economics* 17 (4), 353–373.

Arrow, Kenneth, Bert Brolin, Robert Costanza, Partha Dasgupta, Carl Folke, C.S. Holling, Bengt-Owe Jansson, Simon Levin, Karl-Göran Mäler, Charles Perrings, and David Pimentel (1995): „Economic growth, carrying capacity, and the environment". *Ecological Economics* 15, 91–95. (Reprinted from *Science* 268, 520–521.)

Asafu-Adjaye, John (1998): „An Empirical Test of the Environmental Kuznets U-Curve Hypothesis". *Indian Journal of Quantitative Economics* 13 (2), 67–91.

Auffhammer, Maximilian, Richard T. Carson, Teresa Garín-Muñoz, and Changhua S. Rich (2001a): „Exploring Structural Differences in CO_2 Emissions of China's Provinces". *Paper presented at the EAERE Conference, Southampton, UK, June 2001.* Downloadable at: http://www.soton.ac.uk/~eaere/conf2001/conf2001.html

Auffhammer, Maximilian, Richard T. Carson, and Teresa Garín-Muñoz (2001b): „Forecasting Chinese CO_2 Emissions: A Province Level Approach". *Department of Economics, University of California, San Diego, mimeo.*

Ayres, Robert U. (1995): „Economic growth: politically necessary but *not* environmentally friendly". *Ecological Economics* 15, 97–99.

Ayres, Robert U. (1997): „The Kuznets curve and the life cycle analogy". *Structural Change and Economic Dynamics* 8, 413–426.

Azomahou, Théophile and Nguyen van Phu (2001): „Economic Growth and CO_2 Emissions: A Nonparametric Approach". *Paper presented at the EAERE Conference, Southampton, UK, June 2001.* Downloadable at: http://www.soton.ac.uk/~eaere/conf2001/conf2001.html

Baldwin, Richard (1995): „Does sustainability require growth?". In: I. Goldin and L. A. Winters (eds.): *„The economics of sustainable development"*, New York: Cambridge University Press, 51–78.

Barbier, Edward B. (1997): „Introduction to the environmental Kuznets curve special issue". *Environment and Development Economics* 2, 369–381.

Barrett, Scott and Kathryn Graddy (2000): „Freedom, growth, and the environment". *Environment and Development Economics* 5, 433–456.

Barro, Robert J. and Xavier Sala-i-Martin (1995): *„Economic Growth"*. New York: McGraw-Hill.

Baumgärtner, Stefan and Frank Jöst (2000): „Joint Production, Externalities, and the Regulation of Production Networks". *Environmental and Resource Economics* 16, 229–251.

Beckerman, Wilfred (1992): „Economic Growth and the Environment: Whose Growth? Whose Environment?". *World Development* 20 (4), 481–496.

Beltratti, Andrea (1996): *„Models of Economic Growth with Environmental Assets"*. Dordrecht, Boston, and London: Kluwer Academic Publishers.

Bengochea-Morancho, Aurelia, Francisco Higón-Tamarit, and Inmaculada Martínez-Zarzoso (2001): „Economic Growth and CO_2 Emissions in the European Union". *Environmental and Resource Economics* 19, 165–172.

Birdsall, N. and D. Wheeler (1992): „Trade policy and industrial pollution in Latin America: where are the pollution havens?". In: Patrick Low (ed.): „International Trade and the Environment". *World Bank Discussion Paper* No. 159. Washington, DC: The World Bank, 159–171.

Borghesi, Simone (1999): „The Environmental Kuznets Curve: a Survey of the Literature". *Fondazione ENI Enrico Mattei, Milan, Italy*, Nota di lavoro 85.99.

Borghesi, Simone (2000): „Income Inequality and the Environmental Kuznets Curve". *Fondazione ENI Enrico Mattei, Milan, Italy*, Nota di lavoro 83.2000.

Bovenberg, A. Lans and Sjak Smulders (1995): „Environmental quality and pollution-augmenting technological change in a two-sector endogenous growth model". *Journal of Public Economics* 57, 369–391.

Bovenberg, A. Lans and Sjak Smulders (1996): „Transitional Impacts of Environmental Policy in an Endogenous Growth Model". *International Economic Review* 37 (4), 861–893.

Bradford, David F., Rebecca Schlieckert, and Stephen H. Shore (2000): „The Environmental Kuznets Curve: Exploring a Fresh Specification". *NBER Working paper* No. 8001.

Carlsson, Fredrik and Susanna Lundström (2001): „Political and Economic Freedom and the Environment: The Case of CO_2 Emissions". *Department of Economics, Göteborg University, mimeo.*

Carson, Richard T., Yongil Jeon, and Donald R. McCubbin (1997): „The relationship between air pollution emissions and income: US data". *Environment and Development Economics* 2, 433–450.

Cassou, Steven P. and Stephen F. Hamilton (2000): „The Transition from Dirty to Clean Industries: Optimal Fiscal Policy in a Two-Sector Model of Endogenous Growth". *Kansas State University, mimeo.*

Cavlovic, Therese A., Kenneth H. Baker, Robert P. Berrens, and Kishore Gawande (2000): „A Meta-Analysis of Environmental Kuznets Curve Studies". *Agricultural and Resource Economics Review* 29 (1), 32–42.

Chiang, Alpha C. (1984): „*Fundamental Methods of Mathematical Economics*". 3rd international edition. Auckland: McGraw-Hill.

Chimeli, Ariaster B. and John B. Braden (2001): „Economic Growth and the Dynamics of Environmental Quality". *Paper presented at the EAERE Conference, Southampton, UK, June 2001.* Downloadable at: http://www.soton.ac.uk/~eaere/conf2001/conf2001.html

Cole, Matthew A. (1999): „Limits to Growth, Sustainable Development and Environmental Kuznets Curves: An Examination of the Environmental Impact of Economic Development". *Sustainable Development* 7, 87–97.

Cole, Matthew A. (2000a): „Air Pollution and 'Dirty' Industries: How and Why Does the Composition of Manufacturing Output Change with Economic Development". *Environmental and Resource Economics* 17, 109–123.

Cole, Matthew A. (2000b): „*Trade Liberalisation, Economic Growth and the Environment*". Cheltenham, UK and Northampton, MA: Edward Elgar.

Cole, M. A., A. J. Rayner, and J. M. Bates (1997): „The environmental Kuznets curve: an empirical analysis". *Environment and Development Economics* 2, 401–416.

Common, Michael (1995): „Economists don't read *Science*". *Ecological Economics* 15, 101–103.

Congleton, Roger D. (1992): „Political Institutions and Pollution Control". *The Review of Economics and Statistics* 74, 412–421.

Cropper, Maureen and Charles Griffiths (1994): „The Interaction of Population Growth and Environmental Quality". *American Economic Review* 84 (2), 250–254.

Dean, Judith M. (1992): „Trade and the Environment: A Survey of the Literature". In: Patrick Low (ed.): „International Trade and the Environment". *World Bank Discussion Paper* No. 159. Washington, DC: The World Bank, 15–28.

Deaton, Angus (1992): „*Understanding Consumption*". Oxford: Clarendon Press.

de Bruyn, Sander M. (1997): „Explaining the environmental Kuznets curve: structural change and international agreements in reducing sulphur emissions". *Environment and Development Economics* 2, 485–503.

de Bruyn, Sander M. (2000): „*Economic Growth and the Environment: An Empirical Analysis*". Economy and Environment Volume 18, Dordrecht, Boston, and London: Kluwer Academic Publishers.

de Bruyn, Sander M. and J.B. Opschoor (1997): „Developments in the throughput – income relationship: theoretical and empirical observations". *Ecological Economics* 20, 255–268.

de Bruyn, Sander M., J.C.J.M. van den Bergh, and J.B. Opschoor (1998): „Economic growth and emissions: reconsidering the empirical basis of environmental Kuznets curves". *Ecological Economics* 25, 161–175.

de Groot, Henri L.F. (1999): „Structural Change, Economic Growth and the Environmental Kuznets Curve: A Theoretical Perspective". *Erasmus University Rotterdam*, OCFEB Research Memorandum No. 9911.

Devlin, R.A. and R.Q. Grafton (1994): „Tradeable Permits, Missing Markets, and Technology". *Environmental and Resource Economics* 4, 171–186.

Dietz, Simon (2000): „Does an environmental Kuznets curve exist for biodiversity?". *Institut für Wirtschaftsforschung, Eidgenössische Technische Hochschule (ETH), Zürich, mimeo.* Downloadable at: http://www.wif.ethz.ch:80/html/WIF/Research/workingpapers/kuznets-kurve.pdf

Dijkgraaf, E. and H.R.J. Vollebergh (1998): „Environmental Kuznets revisited: Time-series versus panel estimation: The CO_2 case". *Erasmus University Rotterdam*, OCFEB Research Memorandum 9806.

Dinda Soumyananda, Dipankor Coondoo, and Manoranjan Pal (2000): „Air quality and economic growth: an empirical study". *Ecological Economics* 34, 409–423.

Eglin, Richard (1995): „Reconciling Free Trade and Environment Concerns: A Multilateral Perspective". In: Thomas Cottier (ed.): „*GATT-Uruguay Round. Neun Beiträge*". Bern: Stämpfli+Cie AG, 123–134.

Ekins, P. (1997): „The Kuznets curve for the environment and economic growth: examining the evidence". *Environment and Planning* A29, 805–830.

Ekins, Paul, Carl Folke, and Robert Costanza (1994): „Trade, environment and development: the issue in perspective". *Ecological Economics* 9, 1–12.

Eriksson, Clas and Joakim Persson (2001): „Inequality and the Environmental Kuznets Curve". *Department of Economics, Swedish University of Agricultural Sciences, Uppsala, mimeo.*

Esty, Daniel C. (1994): „*Greening the GATT. Trade, Environment and the Future*". Institute for International Economics, Washington, DC, London: Longman.

Faber, Malte, Frank Jöst, Reiner Manstetten, Georg Müller-Fürstenberger, and John L.R. Proops (1996): „Linking Ecology and Economy: Joint Production in the Chemical Industry". In: Malte Faber, Reiner Manstetten, and John L.R. Proops (eds.): „*Ecological Economics – Concepts and Methods*", Cheltenham, UK and Brookfield, US: Edward Elgar, chapter 13, 263–278.

Feichtinger, Gustav and Richard F. Hartl (1986): „*Optimale Kontrolle ökonomischer Prozesse*". Berlin and New York: Walter de Gruyter.

Forster, Bruce A. (1973): „Optimal Capital Accumulation in a Polluted Environment". *Southern Economic Journal* 39, 544–547.

Fourastié, Jean (1954): „*Die grosse Hoffnung des zwanzigsten Jahrhunderts*". Köln-Deutz: Bund-Verlag.

Fredriksson Per G. and Noel Gaston (2000): „Ratification of the 1992 climate change convention: What determines legislative delay". *Public Choice* 104, 345–368.

Frey, René L., Elke Staehelin-Witt, and Hansjörg Blöchliger (1991). „*Mit Ökonomie zur Ökologie: Analyse und Lösungen des Umweltproblems aus ökonomischer Sicht*". Basel and Frankfurt: Helbling & Lichtenhahn.

Galeotti, Marzio and Alessandro Lanza (1999a): „Desperately Seeking (Environmental) Kuznets". *Fondazione ENI Enrico Mattei, Milan, Italy*, Nota di lavoro 2.99.

Galeotti, Marzio and Alessandro Lanza (1999b): „Richer and Cleaner? A Study of Carbon Dioxide Emissions in Developing Countries". *Fondazione ENI Enrico Mattei, Milan, Italy*, Nota di lavoro 87.99.

Gallet, Craig A., John A. List, and Jason F. Shogren (1999): „Reconsidering the Emission Savings due to the 1970 Clean Air Act Amendments". *University of Central Florida, mimeo.*

Gangadharan, Lata and Ma. Rebecca Valenzuela (2001): „Interrelationships between income, health and the environment: extending the Environmental Kuznets Curve hypothesis". *Ecological Economics* 36, 513–531.

Gawande, Kishore, Alok K. Bohara, Robert P. Berrens, and Pingo Wang (2000): „Internal migration and the environmental Kuznets curve for US hazardous waste sites". *Ecological Economics* 33, 151–166.

Gawande, Kishore, Robert P. Berrens, and Alok K. Bohara (2001): „A consumption-based theory of the environmental Kuznets Curve". *Ecological Economics* 37, 101–112.

Gradus, Raymond and Sjak Smulders (1993): „The Trade-off Between Environmental Care and Long-term Growth – Pollution in Three Prototype Models". *Journal of Economics* 58 (1), 25–51.

Green, William H. (1997). *„Econometric Analysis"*. 3rd edition. London: Prentice-Hall International.

Grossman, Gene M. (1995): „Pollution and growth: what do we know?". In: I. Goldin and L. A. Winters (eds.): *„The economics of sustainable development"*, New York: Cambridge University Press, 19–46.

Grossman, Gene M. and Alan B. Krueger (1991): „Environmental Impacts of a North American Free Trade Agreement". *Woodrow Wilson School, Princeton University*, Discussion Paper No. 158. Also published in 1993: in Peter M. Garber (ed.): *„The Mexico-U.S. Free Trade Agreement"*, Cambridge, MA and London: The MIT Press, 13–57.

Grossman, Gene M. and Alan B. Krueger (1995): „Economic Growth and the Environment". *Quarterly Journal of Economics* 110, 353–377.

Grossman, Gene M. and Alan B. Krueger (1996): „The inverted-U: what do we know?". *Environment and Development Economics* 1, 119–122.

Grossman, Gene M., Alan B. Krueger, and James Laity (1994): „Determinants of Air Pollution in U.S. Counties". *Woodrow Wilson School, Princeton University*, Discussion Paper No. 169.

Gruhl, Herbert (1992): *„Ein Planet wird geplündert: Die Schreckensbilanz unserer Politik"*. Frankfurt am Main: Fischer Taschenbuch Verlag.

Gruver, Gene W. (1976): „Optimal Investment in Pollution Control Capital in a Neoclassical Growth Context". *Journal of Environmental Economics and Management* 3, 165–177.

Halkos, George E. and Efthymios G. Tsionas (2001): „Environmental Kuznets curves: Baysian evidence from switching regime models". *Energy Economics* 23, 191–210.

Harbaugh, William, Arik Levinson, and David Wilson (2000): „Reexamining the Empirical Evidence for an Environmental Kuznets Curve". *NBER Working paper* No. 7711.

Härdle, Wolfgang (1990): „*Applied nonparametric regression*". Cambridge: Cambridge University Press.

Heil, Mark T. and Thomas M.Selden (2001): „Carbon emissions and economic development: future trajectories based on historical experience". *Environment and Development Economics* 6, 63–83.

Helm, Carsten (1995): „*Sind Freihandel und Umweltschutz vereinbar? Ökologischer Reformbedarf des GATT/WTO-Regimes*". Berlin: Wissenschaftszentrum Berlin für Sozialforschung, Edition Sigma.

Herzog, Howard, Baldur Eliasson, and Olav Kaarstad (2000): „Die Entsorgung von Treibhausgasen". *Spektrum der Wissenschaft* May 2000, 48–56.

Hettige, Hemamala, Robert E.B. Lucas, and David Wheeler (1992): „The Toxic Intensity of Industrial Production: Global Patterns, Trends, and Trade Policy". *American Economic Review* 82, 478–481.

Hettige, Hemamala, Mainul Huq, Sheoli Pargal, and David Wheeler (1996): „Determinants of Pollution Abatement in Developing Countries: Evidence from South and Southeast Asia". *World Development* 24 (12), 1891–1904.

Hettige, Hemamala, Muthukumara Mani, and David Wheeler (2000): „Industrial pollution in economic development: the environmental Kuznets curve revisited". *Journal of Development Economics* 62, 445–476.

Hill, Robert J. and Elisabetta Magnani (2000): „An Exploration of the Conceptual and Empirical Basis of the Environmental Kuznets Curve". *School of Economics, University of New South Wales, Sydney, mimeo.*

Hilton, F. G. Hank and Arik Levinson (1998): „Factoring the Environmental Kuznets Curve: Evidence from Automotive Lead Emissions". *Journal of Environmental Economics and Management* 35, 126–141.

Holtz-Eakin, Douglas and Thomas M. Selden (1992): „Stocking the Fires? CO_2 Emissions and Economic Growth". *NBER Working Paper* No. 4248. Also published in *Journal of Public Economics* 57 (1995), 85–101.

Holzbaur, Ulrich, Maximilian Kolb, and Helmut Rosswag (1996): „*Umwelttechnik und Umweltmanagement: Ein Wegweiser für Studium und Praxis*". Heidelberg, Berlin, and Oxford: Spektrum Akademischer Verlag.

Homburg, Andreas and Ellen Matthies (1998): „*Umweltpsychologie: Umweltkrise, Gesellschaft und Individuum*". Grundlagentexte Psychologie, Weinheim and München: Juventa Verlag.

Houghton, John (1997): „*Globale Erwärmung: Fakten, Gefahren und Lösungswege*". Berlin, Heidelberg, and New York: Springer-Verlag.

Howarth, Richard B., Lee Schipper, Peter A. Duerr, and Steinar Strøm (1991): „Manufacturing energy use in eight OECD countries: Decomposing the impacts of changes in output, industry structure and energy intensity". *Energy Economics* 13, 135–142.

Huesemann, Michael H. (2001): „Can pollution problems be effectively solved by environmental science and technology? An analysis of critical limitations". *Ecological Economics* 37, 271–287.

Hung, Ming-Feng and Daigee Shaw (2000): „Economic Growth and the Environmental Kuznets Curve in Taiwan". *National Cheng-Chi University, mimeo.*

Ihori, Toshihiro (1978): „The Golden Rule and the Role of the Government in a Life Cycle Growth Model". *American Economic Review* 68(3), 389–396.

IPCC (Intergovernmental Panel on Climate Change) (1996): „*Climate Change 1995: The Science of Climate Change*". Cambridge: Cambridge University Press.

Islam, Nazrul (1997): „Income-Environment Relationship: How different is Asia?". *Asian Development Review* 15(1), 18–51.

Jaeger, William K. (1999): „Economic Growth and Environmental Resources". *Economics Department, University of Oregon, mimeo.*

Jaffe, Adam B., Steven R. Peterson, Paul R. Portney, and Robert N. Stavins (1995): „Environmental Regulation and the Competitiveness of U.S. Manufacturing: What Does the Evidence Tell Us?". *Journal of Economic Literature* 33, 132–163.

Jänicke, Martin, Manfred Binder, and Harald Mönch (1997): „'Dirty Industries': Patterns of Change in Industrial Countries". *Environmental and Resource Economics* 9, 467–491.

John, A. and R. Pecchenino (1994): „An Overlapping Generations Model of Growth and the Environment". *Economic Journal* 104, 1393–1410.

John, A., R. Pecchenino, D. Schimmelpfennig, and S. Schreft (1995): „Short-lived agents and the long-lived environment". *Journal of Public Economics* 58, 127–141.

Jones, Larry E. and Rodolfo E. Manuelli (1995): „A positive model of growth and pollution control". *NBER Working paper* No. 5205.

Karr, James R. and Tom Thomas (1996): „Economics, Ecology, and Environmental Quality". *Ecological Applications* 6 (1), 31–32.

Kaufmann, Robert K., Brynhildur Davisdottir, Sophie Garnham, and Peter Pauly (1998): „The determinants of atmospheric SO_2 concentrations: reconsidering the environmental Kuznets curve". *Ecological Economics* 25, 209–220.

Keeler, Emmett, Michael Spence, and Richard Zeckhauser (1971): „The Optimal Control of Pollution". *Journal of Economic Theory* 4, 19–34.

Kelly, David L. (2000): „On Kuznets Curves Arising from Stock Externalities". *Department of Economics, University of Miami, mimeo.*

Komen, Marinus H.C., Shelby Gerking, and Henk Folmer (1997): „Income and environmental R&D: empirical evidence from OECD countries". *Environment and Development Economics* 2, 505–515.

Koop, Gary and Lise Tole (1999): „Is there an environmental Kuznets curve for deforestation?". *Journal of Development Economics* 58, 231–244.

Koopmans, Tjalling C. (1965): „On the Concept of Optimal Economic Growth". In: Pontificia Academia Scientiarum (ed.): *„The Econometric Approach to Development Planning"*, Pontificiae Academiae Scientiarum Scripta Varia 28, Amsterdam: North Holland, 225–287.

Kriström, Bengt and Pere Riera (1996): „Is the Income Elasticity of Environmental Improvements Less Than One?". *Environmental and Resource Economics* 7, 45–55.

Kuznets, Simon (1955): „Economic Growth and Income Inequality". *The American Economic Review* 45, 1–28.

Labson, B. Stephen and Paul L. Crompton (1993): „Common Trends in Economic Activity and Metals Demand: Cointegration and the Intensity of Use Debate". *Journal of Environmental Economics and Management* 25, 147–161.

Labys, Walter C. and Lorna M. Waddell (1989): „Commodity lifecycles in US materials demand". *Resources Policy* 15, 238–252.

Li, Eric A.L. (1989): „An Environmental Cooperation Agreement for the Asia-Pacific Region?". *The Australian Economic Review* 32 (2), 145–156.

Lieb, Christoph M. (2002): „The environmental Kuznets curve and satiation: a simple static model". *Environment and Development Economics* 7, 429–448.

Lim, Jaekyu (1997): „The Effects of Economic Growth on Environmental Quality: Some Empirical Investigation for the Case of South Korea". *Seoul Journal of Economics* 10 (3), 273–292.

Liski, Matti and Anne Toppinen (2001): „Stochastic versus Deterministic Trends in Carbon Dioxide Time Series". *Helsinki School of Economics, mimeo.* Downloadable at: http://www.hkkk.fi/~liski

List, John A. and Craig A. Gallet (1999): „The environmental Kuznets curve: does one size fit all?". *Ecological Economics* 31, 409–423.

List, John A. and Shelby Gerking (2000): „Regulatory Federalism and Environmental Protection in the United States". *Journal of Regional Science* 40 (3), 453–471.

Liu, David H.F. and Béla G. Lipták (2000): „*Air pollution*". Boca Raton, London, and New York: Lewis publishers.

López, Ramón (1992): „The Environment as a Factor of Production: The Economic Growth and Trade Policy Linkages". In: Patrick Low (ed.): „International Trade and the Environment". *World Bank Discussion Paper* No. 159. Washington, DC: The World Bank, 137–156.

López, Ramón (1994): „The Environment as a Factor of Production: The Effects of Economic Growth and Trade Liberalization". *Journal of Environmental Economics and Management* 27, 163–184.

López, Ramón and Siddhartha Mitra (2000): „Corruption, Pollution, and the Kuznets Environment Curve". *Journal of Environmental Economics and Management* 40, 137–150.

Low, Patrick (1992): „International Trade and the Environment: An Overview". In: Patrick Low (ed.): „International Trade and the Environment". *World Bank Discussion Paper* No. 159. Washington, DC: The World Bank, 1–14.

Low, Patrick and Alexander Yeats (1992): „Do 'Dirty' Industries Migrate?". In: Patrick Low (ed.): „International Trade and the Environment". *World Bank Discussion Paper* No. 159. Washington, DC: The World Bank, 89–103.

Lucas, Robert E.B., David Wheeler, and Hemamala Hettige (1992): „Economic Development, Environmental Regulation and the International Migration of Toxic Industrial Pollution: 1960–1988". In: Patrick Low (ed.): „International Trade and the Environment". *World Bank Discussion Paper* No. 159. Washington, DC: The World Bank, 67–86.

Lugo, Ariel E. (1996): „Caribbean island landscapes: indicators of the effects of economic growth on the region". *Environment and Development Economics* 1, 128–136.

Magnani, Elisabetta (2000a): „The Environmental Kuznets Curve, environmental protection policy and income distribution". *Ecological Economics* 32, 431–443.

Magnani, Elisabetta (2000b): „The Environmental Kuznets Curve: Development Path or Policy Result?". *School of Economics, University of New South Wales, Sydney, mimeo.*

Mas-Colell, Andreu, Michael D. Whinston, and Jerry R. Green (1995): „*Microeconomic Theory*". New York and Oxford: Oxford University Press.

McConnell, Kenneth E. (1997): „Income and the demand for environmental quality". *Environment and Development Economics* 2, 383–399.

Millimet, Daniel L. (2000): „Pollution and Development: Dissecting the Environmental Kuznets Curve". *Department of Economics, Southern Metho-*

dist University, Dallas, mimeo. Downloadable at: http://www.smu.edu/~millimet/index.htm

Millimet, Daniel L. and Thanasis Stengos (2000): „A Semiparametric Approach to Modelling the Environmental Kuznets Curve Across US States". *Department of Economics, Southern Methodist University, Dallas, mimeo.* Downloadable at: http://www.smu.edu/~millimet/index.htm

Millimet, Daniel L., John A. List, and Thanasis Stengos (2000): „The Environmental Kuznets Curve: Real Progress or Misspecified Models?". *Department of Economics, Southern Methodist University, Dallas, mimeo.* Downloadable at: http://www.smu.edu/~millimet/index.htm

Minliang, Zhou, Cees A. Withagen, and Henri L.F. de Groot (2001): „Dynamics of China's Regional Development and Pollution: An investigation into the existence of an Environmental Kuznets Curve". *Paper presented at the EAERE Conference, Southampton, UK, June 2001.* Downloadable at: http://www.soton.ac.uk/~eaere/conf2001/conf2001.html

Moomaw, William R. and Gregory C. Unruh (1997): „Are environmental Kuznets curves misleading us? The case of CO_2 emissions". *Environment and Development Economics* 2, 451–463.

Moslener, Ulf and Till Requate (2001): „Optimal Abatement Strategies for Various Interacting Greenhouse Gases – Is the Global Warming Potential a Useful Indicator?". *Universität Heidelberg, Department of Economics,* Discussion Paper No. 360. Downloadable at: http://www.uni-heidelberg.de/institute/fak18/index_d.html

Munasinghe, Mohan (1995): „Making economic growth more sustainable". *Ecological Economics* 15, 121–124.

Munasinghe, Mohan (1999): „Is environmental degradation an inevitable consequence of economic growth: tunneling through the environmental Kuznets curve". *Ecological Economics* 29, 89–109.

Murdoch, James C., Todd Sandler, and Keith Sargent (1997): „A Tale of Two Collectives: Sulphur versus Nitrogen Oxides Emission Reduction in Europe". *Economica* 64, 281–301.

Nentwig, Wolfgang (1995). *„Humanökologie: Fakten – Argumente – Ausblicke".* Berlin, Heidelberg, and New York: Springer.

Neumayer, Eric (1998): „Is Economic Growth the Environment's Best Friend?". *Zeitschrift für Umweltpolitik und Umweltrecht* 21 (2), 161–176.

Nordhaus, William D. (1991): „To slow or not to slow: The economics of the greenhouse effect". *The Economic Journal* 101, 920–937.

Opschoor, Hans (2000): „The ecological footprint: measuring rod or metaphor?". *Ecological Economics* 32, 363–365.

Orians, Gordon H. (1996): „Economic Growth, the Environment, and Ethics". *Ecological Applications* 6 (1), 26–27.

Panayotou, Theodore (1995): „Environmental Degradation at Different Stages of Economic Development". In: Iftikhar Ahmed and Jacobus A. Doeleman (Eds.): *„Beyond Rio: The Environmental Crisis and Sustainable Livelihoods in the Third World"*. ILO Study Series, New York: St.Martin's Press, 13–36. Also published in 1993 as Working paper WP238, Technology and Employment Programme, International Labor Office, Geneva.

Panayotou, Theodore (1997): „Demystifying the environmental Kuznets curve: turning a black box into a policy tool". *Environment and Development Economics* 2, 465–484.

Pearson, Peter J.G. (1994): „Energy, Externalities and Environmental Quality: Will Development Cure the Ills It Creates?". *Energy Studies Review* 6 (3), 199–216.

Perman, Roger and David I. Stern, (1999): „The Environmental Kuznets Curve: Implications of Non-Stationarity". *The Australian National University, Canberra, Centre for Resource and Environmental Studies*, Working Papers in Ecological Economics No. 9901.

Perrings, Charles and Alberto Ansuategi (2000): „Sustainability, growth and development". *Journal of Economic Studies* 27 (1/2), 19–54.

Pezzey, John (1989): „Economic Analysis of Sustainable Growth and Sustainable Development". *The World Bank, Washington, DC*, Environment Department Working Paper No. 15.

Ringquist, Evan J. (1993): „Testing Theories of State Policy-Making: The Case of Air Quality Regulation". *American Politics Quarterly* 21 (3), 320–342.

Roberts, J. Timmons and Peter E. Grimes (1997): „Carbon Intensity and Economic Development 1962–91: A Brief Exploration of the Environmental Kuznets Curve". *World Development* 25 (2), 191–198.

Rock, Michael T. (1996): „Pollution Intensity of GDP and Trade Policy: Can the World Bank Be Wrong?". *World Development* 24 (3), 471–479.

Rothman, Dale S. (1998): „Environmental Kuznets curves – real progress or passing the buck? A case for consumption-based approaches". *Ecological Economics* 25, 177–194.

Rothman, Dale S. and Sander M. de Bruyn (1998): „Probing into the environmental Kuznets curve hypothesis". *Ecological Economics* 25, 143–145.

Saint-Paul, Gilles (1995): „Discussion of 'Pollution and growth: what do we know?'." In: I. Goldin and L. A. Winters (eds.): *„The economics of sustainable development"*, New York: Cambridge University Press, 47–50.

Schindler, D. W. (1996): „The Environment, Carrying Capacity and Economic Growth". *Ecological Applications* 6 (1), 17–19.

Schmalensee, Richard, Thomas M. Stoker, and Ruth A. Judson (1998): „World Carbon Dioxide Emissions: 1950–2050“. *Review of Economics and Statistics* 80, 15–27.

Scruggs, Lyle A. (1998): „Political and economic inequality and the environment“. *Ecological Economics* 26, 259–275.

Selden, Thomas M. and Daqing Song (1994): „Environmental Quality and Development: Is There a Kuznets Curve for Air Pollution Emissions“. *Journal of Environmental Economics and Management* 27, 147–162.

Selden, Thomas M. and Daqing Song (1995): „Neoclassical Growth, the J Curve for Abatement, and the inverted U Curve for Pollution“. *Journal of Environmental Economics and Management* 29, 162–168.

Selden, Thomas M., Anne S. Forrest, and James E. Lockhart (1999): „Analysing the Reductions in U.S. Air Pollution Emissions: 1970 to 1990“. *Land Economics* 75 (1), 1–21.

Sengupta, Ramprasad (1996): „CO_2 emission-income relationship: policy approach for climate control“. *Pacific and Asian Journal of Energy* 7 (2), 207–229.

Setzer, Martin (1998): „*Wirtschaftliche Entwicklung und Energieintensität: Zur Theorie und Empirie der Determinanten der Energieintensität*“. Marburg: Metropolis-Verlag.

Shafik, Nemat (1994): „Economic Development and Environmental Quality: An Econometric Analysis“. *Oxford Economic Papers* 46, 757–773.

Shafik, Nemat and Sushenjit Bandyopadhyay (1992): „Economic Growth and Environmental Quality: Time Series and Cross Country Evidence“. *World Bank Policy Research Working Paper* WPS 904.

Shukla, Vibhooti and Kirit Parikh (1992): „The Environmental Consequences of Urban Growth: Cross-National Perspectives on Economic Development, Air Pollution, and City Size“. *Urban Geography* 13 (5), 422–449.

Skonhoft, Anders and Havard Solem (2001): „Economic growth and land-use changes: the declining amount of wilderness land in Norway“. *Ecological Economics* 37, 289–301.

Smulders, Sjak (1995a): „Environmental Policy and Sustainable Economic Growth: An Endogenous Growth Perspective“. *De Economist* 143 (2), 163–195.

Smulders, Sjak (1995b): „Entropy, Environment, and Endogenous Economic Growth“. *International Tax and Public Finance* 2, 319–340.

Smulders, Sjak (2000): „Economic Growth and Environmental Quality“. In: Henk Folmer and H. Landis Gabel (eds.): *„Principles of Environmental and Resource Economics“*, Second Edition. Cheltenham: Edward Elgar, 602–664.

Smulders, Sjak and Lucas Bretschger (2000): „Explaining Environmental Kuznets Curves: How Pollution Induces Policy and New Technologies“. *Center for Economic Research, Tilburg*, Discussion Paper No. 2000–95.

Smulders, Sjak and Raymond Gradus (1996): „Pollution abatement and long-term growth“. *European Journal of Political Economy* 12, 505–532.

Sprösser, Susanne (1988): *„Wirtschaftswachstum und Umweltschutz: Eine theoretische und empirische Analyse der Zielbeziehungen“*. Ifo-Studien zur Umweltökonomie, München: Ifo-Institut für Wirtschaftsforschung.

Stern, David I. (1998): „Progress on the environmental Kuznets curve?“. *Environment and Development Economics* 3, 173–196.

Stern, David I. (1999): „Attributing Changes in Global Sulfur Emissions“. *The Australian National University, Canberra, Centre for Resource and Environmental Studies*, Working Papers in Ecological Economics No. 9902.

Stern, David I. and Michael S. Common (2001): „Is There an Environmental Kuznets Curve for Sulfur?“. *Journal of Environmental Economics and Management* 41, 162–178.

Stern, David I., Michael S. Common, and Edward B. Barbier (1996): „Economic Growth and Environmental Degradation: The Environmental Kuznets Curve and Sustainable Development“. *World Development* 24, 1151–1160.

Stern, David I., Tony Auld, Michael S. Common, and Kali K. Sanyal (1998): „Is There an Environmental Kuznets Curve for Sulfur?“. *Paper presented at the World Congress on Environmental Economics, Venezia, Italy, June 1998.*

Stern, Nicholas H. (1977): „Welfare Weights and the Elasticity of the Marginal Valuation of Income“. In: M. Artis and R. Nobay (eds.): *„Studies in Modern Economic Analysis: The Proceedings of the Association of University Teachers in Economics Edinburgh 1976“*, Oxford: Blackwell.

Stigliani, William M., Peter Doelman, Wim Salomons, Rainer Schulin, Gera R.B. Smidt, and Sjoerd E.A.T.M. van der Zee (1991): „Chemical Time Bombs: Predicting the Unpredictable“. *Environment* 33 (4), 4–9 and 26–30.

Stokey, Nancy L. (1998): „Are there limits to growth?“. *International Economic Review* 39, 1–31.

Sun, J.W. (1998): „Changes in energy consumption and energy intensity: A complete decomposition model“. *Energy Economics* 20, 85–100.

Suri, Vivek and Duane Chapman (1998): „Economic growth, trade and energy: implications for the environmental Kuznets curve“. *Ecological Economics* 25, 195–208.

Syrquin, Moshe (1988): „Patterns of Structural Change“. In: Hollis Chenery and T.N. Srinivasan (eds.): *„Handbook of Development Economics“*, Volume I, Amsterdam: North Holland, 203–273.

Tahvonen, Olli and Jari Kuuluvainen (1993): „Economic Growth, Pollution, and Renewable Resources“. *Journal of Environmental Economics and Management* 24, 101–118.

Tobey, James A. (1990): „The Effects of Domestic Environmental Policies on Patterns of World Trade: An Empirical Test“. *Kyklos* 43 (2), 191–209.

Torras, Mariano and James K. Boyce (1998): „Income, inequality, and pollution: a reassessment of the environmental Kuznets curve“. *Ecological Economics* 25, 147–160.

Torvanger, Asbjørn (1991): „Manufacturing sector carbon dioxide emissions in nine OECD countries, 1973–87: A Divisia index decomposition to changes in fuel mix, emission coefficients, industry structure, energy intensities and international structure“. *Energy Economics* 13, 168–186.

Unruh, Gregory C. and William R. Moomaw (1998): „An alternative analysis of apparent EKC-type transitions“. *Ecological Economics* 25, 221–229.

van Ewijk, Caspar and Sweder van Wijnbergen (1995): „Can Abatement Overcome the Conflict Between Environment and Economic Growth?“. *De Economist* 143 (2), 197–216.

van Kooten, G. Cornelis and Erwin H. Bulte (2000): „The ecological footprint: useful science or politics?“. *Ecological Economics* 32, 385–389.

Viguier, Laurent (1999): „Emissions of SO_2, NO_x and CO_2 in Transition Economies: Emission Inventories and Divisia Index Analysis“. *The Energy Journal* 20 (2), 59–87.

Vincent, Jeffrey R. (1997): „Testing for environmental Kuznets curves within a developing country“. *Environment and Development Economics* 2, 417–431.

Vogel, Michael P. (1999): *„Environmental Kuznets Curves: A Study on the Economic Theory and Political Economy of Environmental Quality Improvements in the Course of Economic Growth“*. Lecture Notes in Economics and Mathematical Systems 469. Berlin and Heidelberg: Springer.

Wackernagel, Mathis and William Rees (1997a): *„Unser ökologischer Fussabdruck: Wie der Mensch Einfluss auf die Umwelt nimmt“*. Basel, Boston, and Berlin: Birkhäuser Verlag.

Wackernagel, Mathis and William Rees (1997b): „Perceptual and structural barriers to investing in natural capital: Economics from an ecological footprint perspective“. *Ecological Economics* 20, 3–24.

Wackernagel, Mathis, Larry Onisto, Patricia Bello, Alejandro Callejas Linares, Ina Susana López Falfán, Jesus Méndez García, Ana Isabel Suárez Guerrero, and Ma. Guadalupe Suárez Guerrero (1999): „National natural capital accounting with the ecological footprint concept". *Ecological Economics* 29, 375–390.

Wang, Pingo, Alok K. Bohara, Robert P. Berrens, and Kishore Gawande (1998): „A risk-based environmental Kuznets curve for US hazardous waste sites". *Applied Economics Letters* 5, 761–763.

Weil, Philippe (1987): „Love Thy Children: Reflections on the Barro Debt Neutrality Theorem". *Journal of Monetary Economics* 19, 377–391.

WTO (World Trade Organisation) (1999): „*Trade and Environment*". Special Studies 4. Geneva: WTO Publications.

Wu, Pei-Ing (1998): „Economic Development and Environmental Quality: Evidence from Taiwan". *Asian Economic Journal* 12 (4), 395–412.

Zaim, Osman and Fatma Taskin (2000): „A Kuznets Curve in Environmental Efficiency: An Application on OECD Countries". *Environmental and Resource Economics* 17, 21–36.

Zinn, Karl Georg (1987): „Fourastié versus Neoklassik. Nochmal: Die aktuelle Strukturdiskussion im Licht der Dreisektorenthese". *Wirtschaft und Gesellschaft* 13, 271–280.